Isabella

Isabella

by

Arabella Weber

Arabella Weber

THE PENTLAND PRESS LIMITED
EDINBURGH · CAMBRIDGE · DURHAM

First published in 1992 by
The Pentland Press Ltd.
Brockerscliffe
Witton le Wear
Durham

ISBN 1 872795 57 9

Cover design by Geoff Hobbs

Typeset by Spire Origination, Norwich
Printed and bound by Antony Rowe Ltd., Chippenham

DEDICATION

To my Great-Grandmother. In MEMORIAM.

L.A.W.

OVERTURE AND BEGINNERS

Overture and beginners

STRELSAU

Seven o'clock, and a cold night

The city of Strelsau is *en fête*, for it is the night of the first day of the New Year. It is the night of the anniversary of His Majesty Karl Ernst Maria Michael Frederik, Grand Duke of Ruritania, Prince of Slavia, Count of Styrinthia and Illyria, Margrave of Karonia, Lord of the Vldava Mark, Grand Master of the Order of the White Rose of St Michael.

It is the night of the day of his Coronation.

It is also the night of the anniversary of Isabella Weber.

The great bell in the tower of the Cathedral of St Michael announces the hour with as much dexterity as the weight of its years allows. One stroke follows ponderously upon another; the vibrations bustle and hustle each other as if there were not room enough in the leaden air for them to spread themselves before the time they mark has passed. This confusion of audible time fills the Stanislausplatz — pride of the citizens of Strelsau; tumbles against the green-painted doors and the bottle-glass windows of the houses upon its periphery; disturbs the cold half-slumbers of the pigeons perched on the cornices of the Royal Opera House; enters sulkily into the corridors, behind the scenes, and even into the meagre dressing rooms of that imposing structure where it sinks unnoticed in the greater confusion, in silence, in corners.

In the largest dressing room (the room with the legend 'Madame Weber' painted in gold upon its white door, for these matters are perfectly arranged at the Opera at Strelsau) a travelling clock upon the table tinkles out a further seven strokes with a deliberation that is an impertinent challenge to the sage accuracy of St Michael's great bell, and without a pause rushes breathlessly into its set piece. The room, hitherto spotted with a jumble of unrelated sounds, is all at once filled with the divinest Mozart.

It is five minutes past the hour, and a cold night. In signal honour of an anniversary and a Coronation, the Opera is about to begin. It seems that it has been about to begin for a great number of seconds, or minutes, or even hours past, according to the temperament of those who are involved in it. Whenever it does begin, it will nevertheless be with the same sense of urgency that seems to mark the presentation of an opera over the whole of the western world. It will seem as though presenting an opera were the last thing performers or management were thinking of either individually or collectively, and as though assisting at such a production were the last thing the audience were thinking of.

Such hysteria. Such chaos. Such getting things right. Such getting things wrong, and then getting them right again. Such tears and sulks and smiles.

All over the world, at almost this precise time, audiences push and shove for the most advantageous places in the foyers, so that they might see and be seen. At the last possible minute they stream, still chattering like a gaggle of farmyard geese, into the House and along the gangways. In the pit the musicians stumble around the music stands and over each other's feet, fumble with their instruments and proceed to tune up with nerve-wracking caterwauls, shrieks and moans. All over the world the conductor takes his place on the rostrum; he shoots his cuffs, bows to the audience; he is greeted with stupendous applause which his appearance does little to merit; he smiles with insincere sincerity, and turns to face the stage.

All over the world curtains go up with such an unexpected velocity that it seems inconceivable they should ever come down again.

The operas will have begun, at last.

One more opera, one more undying performance in a long sequence of undying performances, according to those sycophants whose critical appreciations fill the columns of musical journals, ending with some bathos: *Madame Weber has never, to our knowledge, sung a wrong note . . . She is always a delight to hear . . . .*

A wry smile twists Madame Isabella Weber's mouth momentarily.

It is Madam's business not to sing a wrong note. It is Madame's business to be a delight to be heard, just as it is the business of the butcher to sell prime steaks, the business of the baker to sell digestible bread, the business of the candlestick maker to demonstrate perfection in his art. So, at the precise moment, exactly on the beat, and with perfect attack, Madame Isabella Weber will demonstrate what perfection may be attained should one have the mind to attain it. To her it is one more performance. To the audience it is one more occasion for unthinking idolatry and adoration. To the critics it is the occasion for the parade of tired adjectives, unsupported by either knowledge or understanding.

His Majesty newly created will sit in his Royal Box listening, perhaps even remembering.

The aristocracy of Strelsau will fill the gap (by invitation only and at shockingly augmented prices) between the stage and the box, expectant, and agog with a curiosity that has never, in all the years, been satisfied.

Madame Weber will pay no attention either to His Majesty or to her audience. She will merely sing. The effect of her singing on others rarely causes her concern. Not at all, in fact, in recent years. To her it is such a little matter.

It is twenty years, almost to the hour, since Isabella Weber last appeared on the stage of the Opera House at Strelsau, and because she has decided to visit once again the House where she made her debut in Strelsau, gossip has been urgent to discover the reason why. Gossip discounted as a reason any desire to demonstrate her art to her adoring public. The cognoscenti of Strelsau, while not knowing too much about the private life of Madame Weber, to their regret, did know a great deal about singers of opera. If one is determined to stay away from the scene of one's early triumphs for a mere matter of twenty years, gossip is of the opinion that one should have sufficient determination to stay away for ever. Such originality of action as that shown by Isabella Weber in appearing in Strelsau has therefore set in motion such a great agitation of mind and tongue in Strelsau Society ever since it was announced that she would sing the role of *Cordelia* at the Royal Opera House on the occasion of the coronation of His Highness Karl Ernst Maria Michael Frederik, Grand Duke of Ruritania, etc., etc., etc.

What the gossips do not know, none of them having husbands or lovers on the inside councils, is that the 'appearance' has been negotiated with that delicacy and finesse usually associated with the approach to royalty with a request to open a bazaar or to launch a ship.

Nor are the gossips aware that Madame Weber's eventual agreement was accompanied by an unyielding demand for the biggest fee ever paid to an opera singer in the history of music, a sum so tremendous that the hair of the Direktor turned completely white during the space of time the negotiations were in progress.

There are some straws in the wind, though identification of the source of the wind that blows them would be difficult. It is widely believed, for example, that Isabella Weber is a daughter of the city of Strelsau. There is no knowing whether this is so or not, but the belief is held so strongly that no one ventures to doubt it. It is just that corroborative evidence is annoyingly undiscoverable. Further, it has been observed by sharp eyes in the short time since she arrived in the city that Madame carries upon her finger no token symbolic of the happy slavery of marriage, yet she is invariably referred to and announced as 'Madame' Weber (for in this city, of all cities, the idea that a virgin could sing so divinely, with such feeling and such supremacy of technique, was not to be thought of for a moment). Finally, a most intriguing point of speculation, there is to be in the cast of the opera *Alma Perdita* a young soprano by the name of Lisa Weber, and this circumstance has the gossips in a flurry of conjecture. Confusion is upon confusion confounded by the remarkable resemblance of both the ladies Weber to a person of, one would have thought, unequivocal rank and position, to wit, His Highness Karl Ernst Maria Michael Frederik, Grand Duke of Ruritania.

Upon consideration of all these things such a delicious wave of anticipation swept through the city that, of course, all invitations to assist at the presentation were accepted and replied to without even a moment's delay. It was in effect a royal command; the dislocation of domestic affairs attendant upon having to dine much earlier than usual; the cancellation of parties to be given by one's self or the attendance at those given by others; the agony of indecision consequent upon the eleventh-hour choice of jewels and fans and gloves; the discomfort of driving through a mist of blown snow in falling temperature were as nothing compared to the possible elucidation of the mystery.

And, because they are at heart good bourgeois, the city is prepared to lay its heart at the feet of the divine Isabella Weber, and to rise at the royal anthem. The warmth of their adoration, their enthusiasm for her art, does not at this moment concern her. She is quite simply sitting before her glass, savouring the solitude and the near silence she demands as a right before the going-up of the curtain. Her face is empty, drained of personality and vitality. There remains only that faint curve of the lips that distinguishes the timeless, irritating femininity of the Mona Lisa.

The face is and has been commented upon freely by many whose use of the word 'beauty' is indicative of their ability to accept that beauty is only skin deep, or, perhaps, a fortuitous arrangement of bone, muscle and flesh pleasing to the eye.

That beauty, together with her talent, has ensnared the envious hearts of many.

The curve of the lips deepen involuntarily as a hundred thoughts, a thousand memories, countless moments of time and life arrange, pass, and re-arrange themselves into a patternless design without beginning or end; glimpses of long-forgotten faces and places; squalidly attractive hotels and theatres in Strelsau, Rome, Vienna, Paris, Strelsau . . . . Her face tightens, momentarily. In heaven's name, why was she here? A new thread appears in the pattern, the scarlet thread of danger among the subdued greens and golds of the past, its flaunting brazen colour compelling the conscience to a distasteful self-examination.

Why have I returned to this place?

Almost she wills the image in the looking glass to give her an answer. But there is no answer. Yet surely there is every answer; it is foolish to think that the joy of times past can be re-lived by the mere wish to re-live it. It is equally foolish to be too curious about those actions and reactions which are now half-remembered history. There is little point now in trying to solve problems which could not be solved so long ago.

Is this really why I returned? she asks herself fearfully. *Did I hope to find the solution to a problem that defied solution*?

But here she was, after having vowed that she would never return to Strelsau, much less sing there, even though at that early date they had dubbed her '*Prima*'.

While Art is long, Memory appears to be short, or at any rate woefully negligent. There is no valid excuse that she can think of now for having consented to appear again in the city, when she could have chosen almost any other city in the western hemisphere in the certain knowledge that she would be given an equally royal reception; there is no reason why she should sing on her birthday when there are so many other days in the long year, and certainly it is not incumbent upon her to sing on the first day of the new year.

Except . . . . that it *was* twenty years ago. Except that it is the birthday of His Highness and his Coronation Day. But what is a Prince? Even, as it now seems, a ruling Prince? In what particular is he different from other men? Is his love more tender, his passion more passionate? Are his declarations more sincere or credible? It seems that the outward trappings must be all.

In honour of this day and its magnificent trappings Madame Isabella Weber has been engaged to sing at the opera, and having agreed to that contract she has automatically made herself responsible for the daily bread of a number of other people, and she cannot deny them. Such a course would be entirely foreign to her nature; she has never denied anyone anything, even when she herself was in dire need of nourishment other than bread.

It is a day of triumph and glory, a day too late.

In that room beyond the looking glass there linger restless spirits; ghosts of times past vibrating still with ambition, love, jealousy, arrogance, envy, hatred . . . . No, scarcely hatred. That would be too violent an emanation. The entities which once gave expression to these faults of the spirit are long since turned to dust; the voices which were charged with the force of these qualities are stilled forever.

The glass is darkly silent at this moment.

Isabella draws out her memory to its sharpest point in an endeavour to pierce the glassy barrier, to recapture, to revivify the tenderness, the passion, the ache of that world that stretches back to the edge of forgetfulness.

It is much too late. Memory is feeble; passion is spent. The colourless ghosts of times past are quiescent, hiding in the corners of Memory.

The travelling clock announces, with an aristocratic clearing of its throat, that it is about to sing that additional part of the Mozartian 'Air' with which it proclaims the passing of the first quarter of the hour, and at the sound of the tinkling tune a curious change occurs. The looking glass becomes a mere looking glass again, and the scene within it merely a room in reverse, a phenomenon greatly in accord with certain laws of the Science of Physics. Such laws would also explain the image of a figure standing silent and grim. Isabella exclaims aloud, and turns in her chair.

'Maria, how you startled me. I did not even hear the door.'

'I dare say that you were far away in your dreams. Building new castles in Spain. Or poking about among the ruins of the old ones.'

'I think you may say poking among the ruins, Maria. It is an interesting but somewhat unsettling occupation, I assure you.'

'That I can imagine, madame.' Maria expresses her opinion dryly. 'There should not be any ruins to poke about among. But there, youth must have its fling. If it gets hurt in the process, who is to blame?'

'No-one is to be blamed, Maria. And it was such a long time ago.'

Isabella stares at the image in the glass. A hair of silver among such a wealth of red-gold; the faintest wrinkles under the eyes, an almost imperceptible tightening of the mouth — what do these matter when set against the aridity of self-knowledge? To understand oneself is to understand others.

'It was such a long time ago, Maria. No-one is to be blamed. The trouble with life is that we must live it. We must each make our own way through the forest. If we are scratched . . . .'

'There are experienced woodmen to lead you through a forest.' Maria rubs the side of her nose reflectively. 'Are there none such to point out the dangers of living to you? No teachers with great experience?'

'No teachers, Maria. Our own experience is the only teacher we have. We learn by experience.'

'If I had as many *schillings* as times I have heard that, madame . . . . Madame means, doubtless, that since we refuse to learn by any other method experience is the name we give to our mistakes.'

'Maria,' Isabella demands suddenly, 'Why are we men and women?'

'It would seem, madame, that God in his wisdom made us so.'

'Where is the evidence for such wisdom, Maria, since all men are fools in one way or another? Certainly they match the women,' she added bitterly.

'Not fools, madame, but children. What they want they must have. When they have got it, they don't want it. And talking of men, there are one or two hanging about hopefully in the corridor, having found their way there by bribery. One or two are young, one or two are without their wives. All are, it seems, of some importance, and all are desirous of inspecting Isabella Weber at closer view.'

She pauses. Her eyes meet those of her mistress across the frontier of the looking glass.

'I cannot think,' she continued, 'that our new-found tolerance in coming to Strelsau stretches so wide as to admit *them*.'

It is a statement rather than a question.

'Under no circumstances are they to be admitted, whoever they may be. This is a theatre, a place of work. As such it does not furnish an opportunity for the frivolity of social occasions.'

Maria grins.

'So, one lesson has been learned at long last,' she rejoins, sarcasm very sharp in her tone. 'We have found, have we, madame, that once a door is opened it is very difficult to close it again. *Gott sei dank* . . . your name will certainly go down in the history of the opera. The *prima* who learned her . . . .'

'It matters very little now, I think, whether the door is actually closed or open,' she replied bitterly. 'But since you and I are both old enough to know what we want . . . .'

'We think we do,' interjects Maria, still grinning.

'And to say that we want it or no, let the door remain closed.'

Isabella throws back her head and stretches her arms.

'Come, Maria. Dress me.'

Her voice has taken on a sudden edge. Maria acknowledges the command with a curtsey. Neither the passage of time nor permitted familiarity have destroyed her basic conception of life: that a mistress is a mistress, and that a servant (however important a friend, guide and counsellor she may be) is a servant. Maria curtseys on entering the room, on leaving, and at times in between.

Nevertheless she is far from being as subservient as the gesture would indicate. She is, as Isabella is wont to declare half in amusement and half in exasperation, at once a treasure, a paragon, and (as many would-be swains have discovered in their time) a dragon. She is also completely human, with a wonderful compassion for the poor and oppressed and a weakness for their oppressors. Particularly she admires aristocratic beauty in silks and satins, and aristocratic handsomeness in evening dress and orders. While she dresses her mistress she babbles of such as are even now within the precincts of the House and the manner of their arrival there. She retails their beauty, their handsomeness, and, with a mouthful of pins, their snobbery and their arrogance.

'Such carriages, Madame, such beautiful carriages, all polished and shining with all the arms of the city upon them. The whole of Strelsau must be assisting at your representation. Madam . . . Such *amateurs* of the opera, they must be . . . Such an occasion it will be . . . Hold still, Madame, I am about to arrange the hair . . . . The Count zu und von Pau-Homburg and his Grafin have already arrived. I saw them just a moment ago. How beautiful she is.' Maria sighs deeply. 'But then, she is old enough to know how to be beautiful.'

Isabella grimaces.

'Maria, you are pulling my hair. And you must not say such things.'

'Not even though they be true, Madame?'

'What is Truth? Did not Pilate ask the same question at a far graver moment? It is perhaps, only the half-truth. Who can really tell?'

Isabella smiles through the glass.

Maria sniffs.

'So, I shall simply say that she is rich enough to know how to be beautiful, and leave it at that. Poor woman. It cannot be her age that attracts the Ambassador of Slavonia, whose carriage was just a few moments behind hers.'

A faint shadow of distaste disturbs Isabella's calm.

'His Excellency the Ambassador,' she muses. 'The gentleman who resembles a pig?'

'The same, Madam,' says Maria. 'The gentleman who, on a certain notorious occasion ate with his hat on.'

Their eyes meet, and they smile together at the remembrance of the occasion.

'He has some success with the ladies, I have heard,' says Maria. 'He is a rich pig.'

'He is a stupid pig,' says Isabella, but without malice.

Certainly he is stupid enough to imagine that the admiration of the crowd gathered about the icy steps of the Opera House is for him personally. So proud is he as he descends from his ornate carriage that he might be the *primo tenore* himself; there is indeed a certain superficial resemblance, especially as regards the girth.

His Excellency's face is somewhat mottled, possibly as a result of injudicious tight-lacing under the resplendent uniform, although the tightness does not prevent him from essaying a little bow to the assembled mob clapping ironically with its chapped hands and huzza-ing with its cracked lips. By the time he has reached the top step his chest has swelled several centimetres, endangering its confining hooks and buttons. He passes into the foyer followed at a discreet pace by his *aide* — a lean-faced fellow saturnine of visage — commenting in grating Slavonian that these are good people, and that his soul is happy because they are good; that he is certainly in the mood to enjoy the imminent musical feast; that his eyes will undoubtedly be refreshed by the sight of the delicious Madam Weber. A magnificent woman. What an ambassadress she would make, do you not agree, Fyodor?

'*Da, da,*' Fyodor agrees mechanically.

'A pity her heart is so cold. She has no human feelings whatever.' He is perhaps referring obliquely to other unsuccessful forays in other cities. 'She is a statue, Fyodor. A statue, a lovely statue.'

He inclines again to the mob before he turns and passes within. A faint ironic cheer follows him. His amorous proclivities are not unknown in Strelsau.

From all points of the city the jingle of bells grows louder as carriage after carriage approaches the Opera House, each drawn by a pride of horses. The carriages roll to a stop over the crackling snow, and from them descend the fairest of the city bravely and handsomely accompanied. From belvedere and castle they come to add to the brilliance of the evening; from aristocratic town house and not-too-distant country lodge, all compelled by the necessity to pay homage to their two most cherished possessions, their king and, by no means second, the owner of the most glorious voice in the world.

The voice has been too long absent from its native land; only rumours, brought home by those rich enough to travel, have reached the ears of Madame Weber's countrymen and women; rumours from barbarous places beyond the frontier, from Russia, Austria, Germany and Italy. Even from England, whose people do not understand music, but who profess to enjoy the noise made by singers, especially if accompanied by a brass band.

Tonight the voice, and its owner, belong to Strelsau. It is to be heard before them, and the pampered beauties of Strelsau and their escorts intend to demonstrate, with as little condescension as possible, the depths of their appreciation of their property.

And of course the depths of their loyalty to their king . . . .

Not entirely unaware of this refined hysteria, but maintaining a discreet silence on the subject, Maria arranges the final folds of her mistress's gown. Isabella begins to apply her make-up, working skilfully to create the histrionic mask which is all that her adoring public is allowed to see of its idol. As she works the transformation from Isabella to *Cordelia* is imperceptible.

It is twenty minutes past the hour.

The stage manager knocks discreetly and peeps in. Having ascertained that Madame Weber is indeed present in the flesh — although he has really no need to think otherwise — he smiles and nods at Maria and retires as discreetly as he came.

Her make-up completed, Isabella still sits with closed eyes, completely relaxed, before the looking glass. This attitude is one that she has taught herself to assume immediately before a performance. When the time comes she will sing. Then she will stop singing, and everything will go on just as before. This evening it is especially necessary to have complete control over her feelings and actions, to comport herself as if nothing untoward had ever happened.

She will sing, and depart.

In the silence of the dressing room Maria creeps about her duties, merely casting a glance at her mistress from time to time.

The loudest sound in the room is the ticking of the clock, still busily cutting off seconds from the flowing ribbon of time.

At the front of the House carriages arrive and depart with a regularity as monotonous as that of the tick of the clock. Each carriage takes its place under the great portico as if it were a piece in a game manipulated by supra-human fingers; it sheds its burden of beauty and arrogance, which looks down its nose for a moment, then mounts the steps. The carriage is whipped away again.

All is perfectly co-ordinated, exquisitely timed.

First, of course, and with due regard for the inverted precedence which seems to govern all such affairs (and with the exception of His Excellency the Ambassador of Slavonia) come certain of the professional and commercial aristocracy of the city; solid citizens under whose patronage and approval the Opera House continues to flourish. Even in romantic Strelsau, as elsewhere, the upper middle class is the backbone of the Arts. A box at the Opera, a picture on the wall, a small but very select library all represent an accolade denied to lesser mortals. In possession of these benefits it is a comforting thought that one need not be too closely acquainted with either the physical or spiritual struggles of the artists concerned.

Of these cultural ephemera a box at the Opera is by far the most important. At the Opera one can see and be seen, both situations imperative in the art of keeping one's foot in the correct place on the slippery social slope.

Now come equipages, rather than mere carriages. They shine with paint and polish, they are emblazoned with arms which display progressively an increasing number of quarterings, the horses which draw them are prouder and more mettlesome, and the lackeys are at once more disdainful and subservient. When the occupants of the equipages descend, it is to be seen that the bosoms are whiter, the jewels that lie upon them more brilliant, the gentlemen who attend them more attentive.

Eventually they too yield pride of place. They progress in extended procession, somewhat negligently up the steps; they pass between the trumpeters and the footmen, and are swallowed up in the temporary anonymity of gyrating humanity within the foyer. When next they appear, they are framed in the gold and velvet boxes. From this point of vantage they look down and around at the dapper uniforms, the sophisticated gowns, the elegant waistcoats which swirl, separate, and coalesce like flowers in a garden. Jewels sparkle more brilliantly in the lights of the auditorium; flowers are still virginal in the corsages of those no longer so.

Still the flowers and the jewels come.

At the top of the steps stands the double row of footmen, clothes in white, maroon, and gold; before them the trumpeters in tabards of cloth of gold, who raise their instruments and give air to a short fanfare. The crowd stirs, stamps its frozen feet, blows upon its fingers, raises its cracked voice in a hesitant cheer.

Another carriage.

The Envoy Extraordinary and Minister Plenipotentiary of the Kingdom of Great Britain and Ireland and his Lady, displaying just the right mixture of aloofness and condescension.

'O-oh,' sigh all the girls at the front of the crowd. 'So handsome.' Involuntarily they sketch a curtsey which the Minister, catching out of the corner of his eye, acknowledges with a little bow and a smile before passing on up the steps. His wife walks by his side; she maintains a British face.

The Envoy Extraordinary and Minister Plenipotentiary of the Republic of France, followed by his lady.

'Aah!' sighs the crowd, its eyes wide on beholding such Parisian splendour.

The Envoy Extraordinary and Minister Plenipotentiary of the Court of St. Petersburg. Without his lady.

The Envoy Extraordinary of the Kingdom of Austria-Hungary, who has no lady as yet, but is always hopeful.

. . . Of Germany.

. . . Of Italy and Sicily.

. . . Of the Kingdom of Spain.

It is five and twenty minutes past the hour.

So much splendour can pall. Expectancy gives way to restlessness. Whispers pass to and fro, barbed arrows of supposition, suggestion, possibility. But while speculation mounts and increases, a rider in royal livery mounted on a white horse regards the animated scene at the front of the House with some curiosity for a while before passing round to the stage door, where, in answer to the official request to state the nature of his business, he hands over a large white envelope, and bending low, whispers into the doorkeeper's ear.

The effect of this is rather startling.

'*Ja, ja, jawohl, mein Herr* . . . immediately.'

With a smile on his lips the messenger watches the old bones shuffle along the passage. It is his opinion that they have not moved so quickly in a long while. Then, still smiling, both he and his horse vanish in a great flurry into the darkness and the blown snow.

'Well,' Maria demands of the knock on the door. 'What mouse is seeking the cheese tonight?'

'It is I, Hans.' The feeble voice is scarcely heard in competition with the continuous preparatory noises from the corridor and beyond.

Maria opens the door a bare centimetre and directs a baleful eye through the aperture. She perceives that it is indeed Hans, bent and rheumaticky Hans who is as much a part of the Opera House as the gilded cherubs on the ceiling. She perceives also that he is holding a large white envelope. Her heart beats quickly for her mistress as she remarks the large purple seal.

'For Madame,' says Hans, holding out the packet.

'I will take it, Hans.'

'No, no. It is I who must give it into Madame's own hands. He said so.'

He draws back from Maria's reach.

'Oh, well, if it must be so.'

She opens the door a little wider so that he may enter, but she is quite ready to close it forcefully in the face of anyone who attempts improper entry.

Hans shuffles towards Isabella, holding the envelope in such a manner as to indicate that he wants no part of the business at all. He stares all round the room in an ecstasy of embarrassment.

'Pretty, aren't they?' It is as if he is speaking personally to the roses which fill it to every corner. '*Very pretty*. But we know where they have come from, don't we?'

'It is little business of yours,' says Maria sharply. 'The envelope, goose. Give it to Madame.'

Isabella smiles at Hans. He is rather a favourite of hers, in spite of his uncompromising manner.

'Come, Hans,' she says softly. 'We are women, and therefore curious. What is the envelope? When did it come? How did it come? From whom does it come? Tell me at once.'

'I should want four tongues to tell you at once. I will tell you with one tongue. It is a message from beyond, Madame.'

'A message from . . . . beyond, Hans?'

'You make fun of us, Hans.' Maria's voice is sharper still.

Hans scratches his head.

'Indeed, no, Madame,' he addresses Isabella. 'That is what he said, Madame. A message from beyond.'

'Who said that, dolt? Explain yourself.'

'Peace, Maria. Hans will tell us all. Well, Hans?'

'It was a fellow on a white horse. At the artist's entrance, Madame. He gave me the envelope. He said I was to give it into your own hand, and to say that it was a message from beyond. He said there would be no answer. Then he rode away.'

'No other message?'

'No other message, Madame.'

Isabella turns the envelope over and over. She examines minutely the superscription; her fingers caress the purple seal.

'After twenty years . . . He has not forgotten.'

It is indeed a Message from Beyond. And how could there be an answer to such a message? A ghost cannot be answered. One does not reply in writing to a memory. Her hands, holding the envelope tightly, tremble, but a small light of happiness shines in her eyes.

'Hans, Maria, he has not forgotten, after all, else why should he have sent this message . . . .?'

'Whatever information it conveys,' says Maria tartly.

'He has remembered, after twenty years,' Isabella whispers.

'So much is apparent,' says Hans, without tempering his tone.

He stares at the envelope in Isabella's hands as if by sheer willpower he could cause it to be consumed by spontaneous combustion.

'But Hans, is it not wonderful? He has remembered.'

Hans is convinced that there are tears poised in her eyes, and that they will fall any moment. The foolishness of women. This is not to be borne.

'There is too much gathering of old threads,' he grumbles, and shuffles off about his duties. 'They will not repair a new garment.'

'So,' says Maria. 'We are remembered, are we? It is now quite clear why we were poking about among the ruins. In spite of our conviction that he would not remember, nor be allowed to remember, we were hoping against hope that he would.'

Isabella closes her eyes tightly. The tears no longer threaten. She feels a little ashamed of her intransigent, almost vengeful attitude, and of her insistence that on this evening of all evenings she shall sing *Alma Perdita*. Just as she had sung it on that dire evening twenty years ago.

They had made their own suggestions, as they had the right so to do, of course.

Madame Isabella Weber, they had written. *In homage to your divine voice and your beautiful presence, may we take the liberty . . . .*

They had not been so sycophantic twenty years ago. And seven years before that, when she had been merely the pupil of Herr Weingartner and a struggling singer, they had described her as a 'backyard hen', on account of her immature voice and her unattractive appearance.

Well, times change, and ideas with them. Now, it seems, they chose to forget what they had thought and said, and the city whose gutters had once been her only playground now offered her its wealth and its worship.

. . . . *We humbly suggest*, they had continued, *that 'La Sonnambula' or perhaps 'Lucia di Lammermoor' might be a suitable piece . . . .*

But since after so long a neglect they had come begging — surely begging was the correct word to describe the situation — Madame Isabella Weber had decided without a second thought on her own choice. She would sing *Alma Perdita.*

They were thrown into immediate confusion. It was impossible, they protested. Did Madame Weber not know? . . . . His Highness . . . . It was the Coronation . . . . Even now some echoes of the past still lingered in the royal corridors. Hands were waved, eyebrows raised. It would begin all over again. Could Madame Weber not . . . .

Madame Weber could not. Fully aware of the confusion which her pronouncement had caused, she merely smiled, took pen and paper, and indited a further letter.

I have many demands on my time and my art, she wrote. *Should you eventually decide that my presence is really necessary in Strelsau on the occasion of the Coronation of His Highness Karl Ernst Maria Michael Frederik, I shall sing, I repeat,* Alma Perdita. *Furthermore, I shall require the assistance of Signor Rufino Andelli as Dirigeant, as before. I shall also require the support of the same Principals as on the previous occasion, should they be willing.*

The Committee was shocked to its corporate heart, but there was no choice. It was Madame Weber or none. With the greatest reluctance, and at the greatest expense, Signor Andelli and the Principals, with one exception, were located and, at the mere further cost of a lawsuit or two consequent on the breaking of a contract, were engaged. The Opera House was redecorated from top to bottom. A large number of pieces of scenery were especially reconstructed according to the original designs discovered in a stage carpenter's dusty box. Programmes were set in hand, and special invitations in silver with purple borders. Some further details of Madame Weber's biography and musical education were humbly requested. This request was refused point blank. She was, she informed them, a singer, not an exhibit in a zoo which must have its lineage hanging on a tree for all to read and comment upon.

Furthermore, she added tartly, should any lines concerning her as a person appear in the programme, or for that matter in the Strelsau public prints, she would consider herself at liberty to account the contract terminated.

Completely in her power, they had no alternative but to comply with these conditions. The agreement which reached her by post through the London fog expressed no further sycophancies or pleadings, but a brief request that she should take every precaution to safeguard her health from infection and her person from accident. A last note was added: the insurers had agreed to underwrite the production with the greatest reluctance in view of the extremely aggravated cost.

'They shall have the value of their money,' she thought bitterly. 'I shall sing though I die.'

But to Maria, in their intimacy, she exulted: 'How I shall sing! I will leave them an unforgettable legacy of sound. I will present them with a control, a dignity, a perfection of singing such as they have never heard nor will ever forget. They shall hear what their backyard hen can do in the way of singing. And this time. . . .' She clenched her fists. 'This time, my heart will not be broken.'

Maria's heart was heavy, nevertheless.

'This is unlike you, Madame. Such bitterness. This is not Madame Weber. This is not Isabella. Consider . . . .'

'Did I not suffer, Maria? Have I not suffered for almost twenty years from thoughts and considerations? Did they consider me? Am I not flesh and blood, Maria?'

'Calmly, Madame, calmly.'

'You have suffered with me, Maria, exiled these many years from your own country, distanced from your family, your friends. Yet you speak of calmness.'

'They would do you honour, Madame. They have great respect for you and your art.'

'And they permit me, nay, *invite* me, to sing in Strelsau, twenty years after. It is not to be borne. Yet I will sing, so that he may feel as I sing a little of what I have suffered, that he may feel perhaps some remorse, at least some twinge of conscience.'

*

The envelope crackles in her twisting hands. Her closed eyes are hot with tears she must not shed. She feels no stirring within her at the hour of her triumph, no overwhelming emotion, no life. The little flame of her triumph has died almost before it has begun to burn. Nothing is real except the dry texture of the paper and the impression of the seal under her fingers. Nothing is real except the memories brought back by the heavy perfume of the roses, forced to attain a hot-house maturity in order that they may grace her dressing room for a few all too short hours.

She opens her eyes. She makes herself return to reality. The letter, if such it be, must be read, she supposes. With a calmness she does not know she could achieve, she breaks the seal and unfolds the protesting paper.

Isabella, she reads, *Our birthday, Our Anniversary.*

There is no signature. The message from beyond is reminder enough.

Isabella essays a smile. Almost she laughs aloud. Maria, her homely features reflecting her mistress's reactions to this message from beyond, smiles in nervous sympathy. Whatever the message is, thinks Maria, it is too late.

Our birthday, Our Anniversary. That is all. With what purpose? A letter as superficial as this is indiscretion enough at this time. He has been discreet for twenty years; why has he broken now his self-imposed silence in such a childish melodramatic manner?

Because of a thoughtless fancy, no more than that, surely, there is now every possibility that she will be nervous, unable to sing. Men were ever selfish, she reflects, even at their most unselfish.

Her reflection stares at her through the glass wall of the mirror. The ghosts of memory giggle and sob and sigh. Selfish, they giggle. She *thinks* him selfish.

They collapse in silent laughter and vanish; she is left with her own reflection.

It is thirty minutes past the hour, and a cold night.

With a sigh Isabella folds the letter and puts it and sentimentality away in a jewel box.

A sudden coldness tells that the door is open. She relaxes and takes a deep breath. Her features soften under the shell of the make-up. Once again she is the poised, accomplished *prima donna* on the point of taking the stage.

'Christopher, lovely to see you. And Lisa. This is quite wrong of you, darling, and unfair to the others.'

Lisa pirouettes towards Isabella, falls to her knees, and pouts as she puts her arms about Isabella's waist.

'I could not resist the urge to visit you . . . . such an important occasion, is it not? You look divine, Maman. Simply divine, and I love you.'

'Lisa,' Isabella checks her. 'You really must remember . . . .'

Maria takes the girl's hand and draws her away. '*Madame Isabella Weber*, child! Tonight we are *the Prima*, the first lady of the Opera. The reason for a little reticence is good enough, should you learn of it later. Saint Cecilia herself could not quarrel with it for a reason.'

Lisa withdraws a little, confused and embarrassed. She makes a little curtsey.

'I beg your pardon, Madame. My dress is so beautiful. I am so happy, and so nervous. But I am quite ready, you see,' she says proudly. 'I am trembling only a little. It will pass.'

She holds out her hands. Her fingers have only the slightest suspicion of movement, but the muscles of her face are tight, and her eyes are too bright. Unable to contain herself, she falls at Isabella's feet, clasping her about the knees.

'You will be wonderful, Madame, as always. It is I who may fail. I could not bear that.'

Isabella caresses the shining copper-coloured hair. Her voice embraces the girl.

'You exaggerate, Lisa. Time and experience will correct these supposed faults. Have faith in yourself and your undoubted gifts. You sing well enough, of that there is no doubt. Tonight you will listen to me from the point of view of the drama, not critically, nor in appreciation. You will attend to your cues, your movements, your expression. Tonight you are not Lisa, you are Marella, a village maiden, a figment of the imagination come temporarily to life. And I . . . . I am *L'Alma Perdita . . . .*'

Now Lisa trembles indeed, and clutches Christopher's arm in sudden fear.

'Madame, are you well?'

'There, child, certainly I am well.'

But even as Isabella speaks, her voice seems to come with an effort from a long distance, and the room seems to have the appearance, for a very short time, of revolving about her. Her hand upon her breast cannot prevent the sudden needle-sharp pain from piercing her through.

'It is nothing, child. I suppose I too am a little excited. Christopher, I have had a message, from Beyond, as Hans described it.'

'Isabella, at such a time!'

'Ever thoughtful, or thoughtless, whichever you choose.' She smiles, but there is no happiness in the smile. 'One can scarcely turn away messages from Beyond, Christopher. It is enough that no answer was expected. And you may take heart that none is intended.'

'This is a complication we had not expected. What are we to do, Isabella?'

'We are to do nothing, Christopher, except what we came to do, and then go. I had really meant to do you a kindness by informing you of this message, that you may be prepared, and to show you how little it interests me. I need not have told you. And there is certainly no complication.'

The room has recovered its equilibrium. The mirror stands in its place, its glassy rectangle bright with present reflections shutting out the dark memories of the past. The shadow has passed. There is nothing more to fear for the moment, save the tenseness of the remaining seconds before the curtain rises and the first note is sung. The initiation over, familiarity will envelop her like a warm cloak, the ghostly impersonal faces beyond the footlight will dissolve in the darkness, and the music will flow around her, a strong barrier to the crudities of the outer world.

'Lisa, child, go back to your room now. Sit down, close your eyes, and take a deep breath. Forget everyone, think of nothing. When you make your entrance, sing. That is all.'

'Very well, Madame. I will do as you say.'

At the door she half turns. Somewhat wistfully she asks:

'But do you not think my dress is beautiful? Christopher says it is too beautiful to be the dress of a maidservant. You do not think so, Madame?'

'You are the vainest creature, I declare, Lisa. The dress is beautifully adequate. It is after all nothing but a *costume de theatre*, nevertheless there may be the possibility of a number of broken hearts in the audience tonight. And now, child, go. The time draws near.'

'Madame, you are an angel.' She pirouettes and turns to the door. 'Oh, Signor Andelli.'

Her eyes twinkle impishly. Standing on tiptoe she kisses him, leaving the print of her lips on his ascetic cheek.

'Pfui!' he exclaims. 'Lisa, Lisa.'

'I love you, I think,' says Lisa. 'It is a fault that you are so bad-tempered.'

'Lisa, child,' Isabella reproves her. 'She is still a child, Rufino.'

'Time will correct that, Madame.'

Lisa closes the door behind her with an exaggerated flourish.

'But I am not bad-tempered,' Signor Andelli protested to the room in general. 'I am never bad-tempered. Well, hardly ever. Madame, there are but twelve minutes remaining. It is precisely,' he consults his watch, 'It is precisely thirty-three minutes past the hour, and he, in whose honour . . . . He has not yet arrived. I am ready, you are ready, the orchestra is ready. *Dio mio*! Even the audience is ready. But he, the Grand Duke of Ruritania, he is not ready. It is not to be borne.'

'Unpunctual as ever,' murmurs Isabella.

'Pardon, Madame?'

'Patience, Rufino. There are yet six hundred seconds remaining. Much may happen in that huge span of time.'

Signor Andelli flings back his black mane, and combs it agitatedly with his fingers.

'It is a disgrace,' he storms. 'It is an insult to you.'

'Punctuality is the politeness of kings. Rufino. But while men must be punctual, princes may perhaps delay a little, or be delayed. Who knows all the circumstances? Princes too have their exits and their entrances, and they make the best of them.'

At that exact moment a bell jangles stridently. Rufino freezes in the middle of a gesture intended to consign to perdition all princes, and in fact all who had to do with opera in any way.

'Aha,' he breathes. 'At last it is he. When he has taken his bow, it will be time for important things. Madame, I salute you.'

Barely pausing to kiss her hand, he flings himself through the door in a flurry of coattails and flying hair.

'Now we can be quiet for a little space,' grumbles Maria. 'We might be in the town square instead of in the prima's dressing room.'

Isabella sighs, and turns once again to the looking glass for the last consultation. Beyond her image the reverse stretches into an infinity of darkness where hover the ghosts of the past, the ghosts of twenty years ago, ghosts with a question still not answered: *You torment only yourself, only yourself. Why*?

Isabella stares into the glass, past her own face, past the comforting and comfortable shape of Maria (who has nothing to do now except await events with some foreboding), past the four walls of the room to the end of the shimmering glass to where the velvet-britched flunkeys stand at attention.

The bell rings again urgently.

The trumpeters raise their instruments to their lips to play a fanfare.

The call boy spirals his way down and round about to the heart of the matter.

Overture and beginners . . . .

Messieurs, Mesdames, Overture and beginners, s'il vous plait . . . .

It is seven forty, and a cold night.

*

Outside the crowd surges forward, each individual intent upon obtaining the best possible view of His Majesty as he descends from the coach, followed by his pudding of a Queen. It is not an elegant trick to perform even in the best of circumstances, and Amelia Viktoria's measurements are ill adapted to the awkward doors of coaches. To those observing her antics it would not appear that she had so recently been invested with the dignity of a ruling prince's consort.

His Majesty ascends the steps, carefully brushed clean of snow once more. He acknowledges the crowd with a slight gesture of the hand here, a turn of the head there, and disappears through the fanfare into the golden crimson of the Grand

Foyer. The stentorian voice of the gentleman usher announces his approach to all within. The sound echoes about the foyer, and even penetrates into the auditorium, where it goes round and round the heads of the audience until they are compelled to turn, as flowers turn to the sun, towards the Royal Box.

'His Majesty Karl Ernst Maria . . . Maria . . . Maria . . . .' the gentleman-usher proclaims, and the echo goes round and round.

'Her Majesty Amelia Viktoria Konstanzia . . . ia . . . ia . . . .'

His Majestymajesty . . . .

Her Majestymajesty . . . .

Karlarl . . . .

Ameliaelia . . . .

Karlarlameliaeliaiaia . . . .

Mariaiakonstanziaiaia . . . ia . . . .

The echoes ricochet from pillar to post until, finally weakened, they spiral tenuously upwards towards the bronze filigree of the dome and are lost in the coldness of the night.

His Majesty mounts the Grand Staircase, acknowledging with the same weariness with which he had greeted his people the bows and curtseys on either side. If he feels any emotion he does not now show it. His face is a mask, as devoid of expression as is Madame Weber's in the dressing room.

Up and up he goes, the Opera officials bowing and scraping at varying but respectful distances to the front and side, until they arrive at the gold and white doors of the Royal Box.

Here His Majesty pauses. The mask of his face is pale. Her Majesty by his side pants and visibly perspires in a most unmajestic manner.

White-gloved footmen obsequiously open the doors, exposing a plushness of golden crimson and white. The curtains are drawn apart; in the distance is to be seen the great proscenium arch, carved, decorated, bescrolled, the grinning grimacing masks of Comedy and Tragedy pendant at its apex.

Below it, Andelli stands on the rostrum. He too faces the Box, bows once, and turns about, his baton raised. The gentlemen of the orchestra wait, poised.

His Majesty advances to the front of the Box and stands unmoving, entirely composed. Her Majesty, still panting and perspiring, stands a little behind him, flanked by the Ladies and Gentlemen-in-Waiting who are on duty this evening.

The footmen bow, and close the doors silently. Beyond the closed doors, they stand one on either side.

The buzz of the audience in that vast honeycomb is stilled for a moment. At the first notes of the Anthem there is a roar of simple movement as it rises to its feet. His Majesty bows to the distance, once, twice, to left and right, and takes his seat.

The doctor, sitting with the aged Matthild in the lower stage box, smiles ironically as he notes the weariness of the royal face.

'So suffer we all,' he muses. 'Our sins *do* find us out.'

The audience settles itself in happy anticipation. It now has its Prince, its favourite singer, and its (almost) favourite *dirigeant*. In a very little while it will have its favourite music, and the cares of the world will be temporarily forgotten.

It is precisely seven forty-five. Alone in the dressing room the little clock announces to itself the three-quarters of the hour, playing three-quarters of its melody.

The baton rises higher, a warningly poised wing-enfolded bird. It flashes downwards, and the long sweep of the violins falls tremblingly downwards into the dark pool of silence.

*

ACT I:

Before the Church of the Holy Sepulchre. Evening.

STRELSAU

ACT I: Before the Church of the Holy Sepulchre. Evening.

STRELSAU

The audience shivers involuntarily as the opening strings give expression to what has long been recognised and quoted at the most moving cry of despair ever to be written in musical symbols. It is the first theme of the Overture, anticipating that upon which Cordelia makes her first entrance, and its demands upon the technique of the singer are awesome. On stage, coming as it does to ears scarcely attuned to receive it (as if the Trump of Doom were to sound in the middle of a Charity Ball), it has been the cause of bitter disputes throughout the world of music.

Its composer was from the first vigorously condemned by traditionalists as a mountebank, an exhibitionist, an innovator without any understanding of the true spirit of music. From them he learned that it was a bad thing to depart so wilfully from the rules laid down by one's preceptors in the art. The more tolerant of them, of whom there were a few, merely shook their heads at the boldness of Signor Andelli's innovatory ideas, and sadly prophesied that, however shockingly effective it might be as pure music, as opera it would certainly prove to be unsingable.

The passage of time seemed likely to justify their prophecy. After its first public performance (at Strelsau, of all places) conducted by the composer himself, and the title role sung by Fraulein Isabella Weber, the strangeness of its musical ideas, and the demanding disciplines required to sing it, caused it to be omitted from the repertoire of those Houses which, since their funds needed constant replenishing, depended on more familiar works.

It stood in grave danger of becoming a musical curiosity.

This momentous first production, as may be learned from histories of music, was the occasion of the twenty-first anniversary of the birth of His Highness Karl Ernst Maria Frederik, Grand Duke of Ruritania and heir to the Throne.

The Fraulein Isabella Weber was then herself only twenty-one years of age, and fresh from an overwhelming success at the Vienna Staatsoper.

The Grand Duke, as fond of music as he was of those who practised it, was noted for his generous patronage of both. At the conclusion of this particular performance he was moved to express in person his appreciation of the work and his thanks to the lady on whose fair shoulders (his own words) the burden of the magnificent representation had fallen. With this intent in mind he had made his way to her retiring room. Protocol was shattered. Many singers had been presented to His Highness. This was the first time His Highness had been presented (by courtesy of the lady's handmaid, no less) to a singer.

As Maria herself remarked, giggling, even for princes there must be a first time for everything.

The evening was from all points of view a complete success. For some time after this evening, people overhearing some snatch of *Alma Perdita* in a restaurant or *bierhalle* would pause in the conversation, listen pensively, and say: 'I know that tune. What is it called now . . . ? I shall think of it in a minute.' This, as everyone knows, is the height of success.

As for Isabella, she might well have misquoted: *I came, I sang, I conquered.* Even before the opera was well begun those who had come in duty bound (among them the 'Friends of Music' of Strelsau) remained to admire, although they considered that they had every right to criticise the lady, for it was common knowledge that, even at the tender age of twenty-one, her pride was greater than theirs. The knowledge that she had vowed to the Herr Kapellmeister Josef Sigismundus Weingartner that she would never sing in Strelsau was not so common. He had kept that information very close indeed, partly because the vow had been made during sickness and in anger, and partly because she was a woman. Every woman, especially if she is a renowned opera singer, has a price at which her vanity is for sale. He considered that sooner or later a desire to show the 'Friends of Music' that their 'backyard hen' had become, in the opinion of all who knew about singing, a nightingale, would be too strong for her to resist. The anniversary of His Highness the Grand Duke seemed to provide an opportune and justifiable excuse.

It may also have been that she would be able to assuage her curiosity on the spot about a man whose mode of life (seemingly modelled on that of his late uncle) was already a byword in high Central European circles.

The Fraulein Isabella Weber returned to Strelsau. In spite of their prejudice against her pride and arrogance, which were already well known in Strelsau, the good citizens were nevertheless captivated both by her beauty and her singing. Rumour — carried from Vienna — had by no means exaggerated those qualities. From that first evening Isabella Weber became a legend in her own country.

The Overture continues.

The second theme is happier, relaxing the rigor induced by the first; it is at this point that the audience settles down and thinks to give itself over entirely to the music.

In his box, His Majesty listens with a set face. The Overture is as familiar to him as if he had heard it only yesterday. He anticipates in his mind the major chords which announce the end of despair and the conquest of bitterness; the resolution of angry question and answer into the calm certainty of understanding and acceptance; he awaits the golden *fortissimo* that speaks of knowledge and hope, that ends the Overture, and raises the curtain.

It seems now that His Majesty is a little more relaxed than when he entered the House. Yet there is still some disturbance of the heart and mind, the Doctor thinks, taking a last look at the mask in the Royal Box before directing his own attention to the stage and the imminent entrance of Madame Isabella Weber.

Actions in life must have their reactions. Take what you want, said God, and pay thereafter.

The newly elevated Queen, Amelia Viktoria Konstanzia Camilla, sits plumply in her gilt chair, one podgy hand lying limply on the plush, the other gently agitating a fan which does little to dissipate the perspiration moistening her pallid brow. The jewels on her much be-ringed hand sparkle as it moves tiredly to and fro. Twenty years ago, His Majesty reflects bitterly, she had been a stolid, unlovely child. Nothing has changed; she is now a stolid, unlovely woman. The music flows over and about her, engulfs her in a wave of sound the like of which has not been heard since Herr Wagner conceived *Tristan and Isolde*, but Amelia Viktoria sits, just as she sat a few short hours ago during the prolonged coronation ceremony; just as she sits in her bedchamber while her ladies dress and undress her; sits as she always sits through their endeavours to amuse her during the long dreary afternoons. Just as she always sits.

With an expression of irritation His Majesty turns again toward the stage, but its depth are hidden by the great gold and purple curtain. The curtain moves, trembles, and is still again.

The Overture is finished, and the silence has an almost minatory tension. A short buzz of appreciation from the audience dies away.

Amelia Viktoria selects one of her favourite bonbons; the sound of her teeth closing on its fragile succulence is the roaring of a miniature Niagara. The warmth of her plump body wafts about her Lord and Master. He shifts uneasily in his chair.

Amelia Viktoria raises her fan.

Andelli raises his baton.

The curtain rises . . . .

Upon the stage now at last disclosed, the ladies and gentlemen of the chorus emerge severally from the wings. They hold their small breviaries within clasped hands; they walk sedately. They converge upon the door of the church.

The music begins softly but with a marked rhythm. A gradual *crescendo* leads into the humming chorus of the pious villagers. Above the cacophony of the bells it rises and falls hauntingly on a silent audience. The last of the congregation enters the church, from where the sound of an organ overrides the *diminuendo* of the violins and the 'cellos.

A small almost tangible silence.

The baton flashes, and from the violins is heard once more that searing introductory theme, with its breathtaking downward sweep. A pure silvery sound, at first scarcely noticeable in the general weave of orchestration, grows very gradually louder so that it fills every corner of the House; the sound of a voice so pure that it is indistinguishable from the singing of a stringed instrument. It is borne upon a sigh from the audience; a whisper so soft that almost nothing of it can be understood as human speech goes round the auditorium:

'Cordelia . . . .'

'Isabella . . . .'

'Isabella Weber . . . .'

Her beauty is legendary, but the audience in the Royal Opera House of Strelsau is shocked. The ravaged face from which that exquisite voice issues is painful to see; the despair upon it owes nothing to histrionic art, it is real, living. No lessons in 'dramatic art' can have taught her to simulate such unhappiness, such sorrow. Such emotions cannot be 'acted'; they can only be suffered. Such is Isabella Weber's power of communicating her art that each person in the audience feels the great unhappiness personally, and he or she is disturbed by the feeling. They do not wish, individually or collectively, to be burdened with any part of Cordelia's suffering; they tell themselves that it is only play-acting, that Madame Weber is only a woman singing; that in a short time it will all be over. It is in vain. So compelling is the person, so exquisite and beguiling the voice, that they share the agony of that imaginary lost soul against their will. The limelight picks out the figure of Cordelia advancing from a hillock beyond the corner of the church. She is clothed in white, mystic, wonderful. Pride at odds with Despair are evident in every line of her figure. The words of her *aria* are now to be heard clearly and strongly — *Sono pellegrina; son' orfana* . . . . The voice is certainly all that has been rumoured of it; the person is beautiful beyond comparison. The modulation of the voice, its colour, its warmth even in the most despairing of passages, rejoice the spirit. Its precision is perfection itself. Yet such perfection might indeed be a reason for criticism; a jewel without a flaw can be quite boring.

The *aria* ends with a sobbing fall. A sigh of release wells up from the audience, and is carried away on the swelling organ music from within the church. As Cordelia stands motionless on the stage, those who are nearest to her can see the tears pendant upon her cheeks.

'What realism,' they say. 'What an artist she is.'

They do not know that Madame Weber has broken the first rule of her art, that is: never to allow personal emotion to show on stage. The tears are real, unbidden.

'*Du lieber Gott*,' exclaims Maria, agitated in the wings. 'She is weeping, Herr Christopher. Madame is weeping.'

She gripped him tightly by the arm.

'That damned letter,' he mutters angrily. 'High or low, human beings do the stupidest things.'

He feels sick at heart, and quite helpless. What can a mere agent, even though he be a step-brother, do in such circumstances?

'But what are we to do, Herr Christopher?'

'What can we do, Maria? Let us hope she regains some control over herself. Let us hope that she does not breakdown completely. I should never have permitted this performance.'

Even old Hans wipes away a tear with the back of his hand as he stands as close behind them as he dares; his being there at all is contrary to the regulations concerning the duties of a stage-door keeper.

'There has been too much gathering of old threads,' he repeats, wagging his shaggy head at no-one in particular. 'I knew it would come to no good. But *they* like her, and that's all that matters.'

Indeed they do like her. It is a moment they are unlikely to forget, such a moment as comes perhaps once in a lifetime. Isabella Weber, the Superb, the *Assoluta*, singing as she has never sung before, like a bird that sings, weeping, and dies of a heart that is too full.

The ladies and gentlemen of the chorus, now issuing, intoning, from the church glance at each other in wild surmise. The rehearsal had not prepared them for such a tableau. Yet it might be supposed that they are well enough acquainted with temperamental sopranos not to be brought up by anything that may happen on stage. But most of them are still young and have little knowledge of the dustier corners of life. Thus it is that all they really see — and hear — is Madame Weber about her professional business.

Lisa, in her character of Marella, sings her small part to comfort her mistress as best she may. She does well enough, but Cordelia is unmoved, stony-faced. Looking beyond the footlights, beyond the wrapt and silent audience, beyond the sumptuous foyer into the snow-covered Stanislausplatz, Isabella sees a ragged unkempt girl-child selling matches on the very steps of the Opera itself, a child who sometimes sings to attract possible customers. In the babble and loquacity of those who pass by her little voice still rises clear and distinct. 'It is quite amusing,' they say, giving her a penny. 'We get two performances for our money.'

'Why are you crying, little girl?' some asked her as they put a coin into her grubby hand. 'No, keep your matches. It is almost the day of St. Nicholas. You should be happy. You will have presents.'

On stage Cordelia sings, in a minor key, a partial *reprise* of her aria. A chorus of villagers gathers about her, expressing bewilderment at her condition. In a duet with Marella, Cordelia rejects her sympathy and her expressions of love, and complains about the emptiness of her life. The percussionist simulates the sound of the church bell tolling the hour of Vespers upon his instrument. It is followed by a silence, a terrible nerve-wracking silence. The audience waits in an agony of expectation for the next note of sorrow.

The baton rises. It summons up and releases the chord of despair from the shuddering strings, a sound that freezes the heart. Again that voice:

Senz'amore, senza vita . . . . Come la vita e sempre vuota . . . .

Silence falls again at the end of the area. Cordelia sinks half-fainting to the boards.

*

'She must be very cold,' said a pair of sweethearts as they passed by. 'She should not be here on such a cold night. Come. We shall be late.'

The bell in the tower of the Cathedral of St. Michael tolled the half-hour. The sound reverberated about the dark, snow-obscured Stanislausplatz.

It was seven-thirty, and a cold night.

There was a great noise in the child's ears, like the sound of many people clapping. A young man passing lifted her to her feet, and rubbed her frozen hands.

'Who are you, little girl?' he said. 'What is your name?'

'They call me Isabella,' she said.

'Isabella who? Have you not another name? And where do you live?'

'Just Isabella. Sometimes ''Bella''. That is all. I live over there.'

She indicated vaguely some direction beyond the present place.

The young man rubbed her hands once more, and gave her some coins.

'Take this money and buy yourself some coffee,' he said. 'Then perhaps you should go home. It is too cold.'

'I wonder who she is,' said the young man. 'She is really quite pretty.'

'She is always here at this time,' said the young woman, pulling her furs closer about her. 'She sings quite often. I call her ''little Jenny''.'

The two giggled happily at this simple fantasy. They were in love, and the young man pressed his lady's arm affectionately as they went up the steps. They had undoubtedly missed the Overture, but it did not matter since they knew all the music, and even some of the words of the opera by heart.

The tears were cold, and fell like icicles on the child's heart.

She wondered often after that why the young lady had called her 'little Jenny', until much later one of the musicians of the Opera told her, whilst buying her matches, that Jenny was a singer too, Jenny Lind indeed was a famous singer. When she sang, people clapped their hands and gave her money. Jenny Lind was rich, she had fine clothes.

Some people gave the little girl money when she sang, but the important thing about singing was that then she was carried into a kind of Heaven where it was no longer cold, and she no longer tired, where she no longer had to sell matches to gentlemen with gold rings who did not want matches, but who wished to appear generous and sympathetic. In Heaven there were no horrid boys who stole her matches, and even once threw one of her tattered boots into the fountain, and ran off crying: '*Look at our nightingale fishing.*' In Heaven there was nothing of this; only music, and a wonderful warmth.

When she stopped singing, then the vision was at an end, and it was cold again.

In the depth of winter when it was really cold and the snow lay thick upon the streets and the roofs of the houses, and draped the statues with pure white togas, she would sometimes strike some of her matches. They were good matches, made by her father when he was sober from resinous pinewood. They burned a long time, and she warmed her fingers at the bright flame.

But in the end a match burnt up, and she was still cold.

In the evening it seemed to be warmer, for in the square between the Cathedral of St. Michael and the Opera House there were many bright lights, and many people going to and fro. The statue of the ancient grand Duke Stanislaus in the middle of the square was a sombre, snow-covered companion, but his permanence was comforting. The people who passed in the square laughed and chattered, and the sleigh bells tinkled.

When the lighted facade of the Opera House went dark and the little girl judged that her father would be snoring off his drunken stupor, she would creep up the stairs to the miserable room in the squalid tenement, wrap her frozen limbs into the blanket which was her only covering, and try to sleep. Sometimes there was food, crusts and crumbs, and the lees of a wine bottle. In the morning, rousing herself from the fitful sleep, she crept out again, leaving the money she had been given the day before on the table and taking a fresh stock of matches. Only in this manner was she able to avoid a possible beating.

So the child passed the dark days of winter.

In summer it was better. The sun shone warmly into each corner of the square in turn. The Grand Duke Stanislaus bestrode his prancing steed boldly in the mellifluous air, and the little girl was warm. She was of course still hungry, for when she ate she had only crusts of the cheapest and blackest of black bread, and anything better than water to drink she picked up where she might.

Sometimes, during the warmest of summer months, she did not go home at all, since it was warm enough, and pleasanter, to sleep in a woody clearing outside the city.

She was twelve years old. She might have been any age.

Among those who passed that way frequently, indeed daily, and consequently noticed the child from force of habit rather than by voluntary perception, was an expatriate Englishman. At first glance he appeared old rather than young. He had what is usually called 'a scholar's stoop', and his eyes were weak. This latter fault he corrected by means of an ancient pair of steel-rimmed spectacles which fitted him so badly that he invariably looked over them instead of through them. This gave him an air of quizzical disbelief that many found disconcerting. The principal interest in his rather haphazard life were his music (he played, as he had played for many years, first violin in the Opera Orchestra and knew the scores of nine major operas by heart) and his son (whose mother, originally a native of Strelsau, had died but a few months previously).

He had the tenderness of the best of Englishmen for all small creatures.

Whenever he perceived the child through the abstraction which was his normal state of mind, a kind of pity went through him for the emaciated body in the tattered dress. He rarely bought any matches, having no great use for them, but he occasionally gave her a small silver coin, fumbling shamefacedly in all his pockets, and he would mutter something to prevent her protestation of gratitude. The look in her eyes was thanks enough.

When, for the first time, he chanced to hear her sing, he was taken aback. The memory of the singing remained with him for a long time, and he looked at her thereafter with a new thoughtfulness. It occurred to him finally that her voice, immature though it was, had a quality which was quite out of the ordinary. This idea caused him some anxiety. He began to listen with attention to the voice, for he knew that in the end he would have to do something about it.

The weeks and the months passed. He did nothing, but he always thought of doing something.

Every evening in the season, as he came to the Opera House, his violin case tucked under his arm, it became a habit to seek out the child consciously from the throng about the steps. They only looked at each other for a moment at first. Then they smiled, diffidently. A kind of friendship grew, naturally, innocently. She appeared to accept him eventually as someone she could trust in an untrustworthy world. They exchanged a few words from time to time, and his heart was full. He gave her biscuits, or a packet of sweets. Sometimes she went with him as far as the stage door, saying nothing, but walking with a childish dignity as though she drew some personal pride from the association.

Always, before he left her at the door, he would bid her 'Good evening', gravely, and she would reply with a seriousness beyond her years. After each of these evening encounters the Englishman felt ridiculously light-hearted, and it is certain that he played his music the better for them, although no-one noticed it particularly.

Such an unwonted association could scarcely have gone unnoticed in Strelsau, where many children once brought into the world, are generally left to fend for themselves at an early age. The children who gathered as gregariously and noisily as sparrows at the base of the statue and on the steps of the Opera House were quick to refer to the child as *the Englishman's Isabella*, which was a sort of compliment, although it was not intended as such.

Sometimes they called him *Isabella's Englishman.*

One evening, in the middle of the season — the evening of the first day of the New Year she was not in her usual place when he passed. He discovered, as everyone discovers sooner or later, that something taken for granted only acquires its real value when lost. That evening he played several wrong notes and earned hard looks from the conductor.

After a sleepless night, and a tormented day, he was too weary, and angry with himself to feel relief when he found her the next evening. His heart nevertheless constricted when he perceived her red-rimmed eyes and the dark bruises on her arms and legs.

Something snapped in his mind.

'Child, child, what has happened?'

'I did not sell enough matches. My father beat me,' she said, as if it were a thing of no consequence now that it was over. 'He was drunk.'

'I had no idea, child. Does he beat you often?'

'Not so often now. But it hurt yesterday.'

Mutely he held out his hand, and she took it as though it were the most natural thing in the world. Together they went to the stage door. Here, instead of saying goodnight, he pulled her roughly into the door-keeper's room, and, his spectacles seeming to flash with decision, bade her to sit in the chair by the fire till he came again.

'This young lady is a friend of mine, Hans,' he bravely informed the young man who had watched the invasion of his sanctum with some astonishment.

'So much is apparent,' retorted the young man, raising his cap and scratching his head vigorously. 'So much is apparent. But . . . .'

'She is cold, Hans. She must be warm.'

'Here it is warm enough, but . . . .'

'Not a word, Hans. Not a word.' The Englishman approached Hans in a conspiratorial manner and laid his finger alongside his nose. 'Should anything be said, refer to me immediately. I will make it right.'

'O-ho, that makes a difference,' said Hans. 'That makes a difference, Herr Smith.'

'Hans,' Herr Smith pulled the young man's head down to the level of his own. 'She *sings*.'

Hans threw up his hands in mock horror.

'*Himmel*. Another. My friend, they all sing. That is the worst of it. If only they would not sing sometimes. It only needs a little will-power . . . .'

'She sings like an angel.'

'She scarcely looks like an angel.'

'She is not like the others, Hans. She *sings*, I tell you. Take the greatest care of her, Hans, until I return. Sit there, child, *zum wiedersehen* . . . . You will be quite warm there. I must go now.'

But she continued to shiver in this strange room. She clutched her bundles of matches in her tattered dress in fear. She would have to confess that she had sold none, and once more she would be beaten. Tears welled up, obscuring the outlines of the room, the table, the pigeonholes, the portraits pinned up at random on the rough walls. The tinkle of a spoon in a cup was the pleasantest of sounds. Hans was making coffee. From across the square the heavy tolling of the bell in the cathedral tower came to the child's ears, mingling with the sound of the coming and going of many feet, the murmur of muted voices, a singing voice practising a bar or two in the distance. It was all so strange, and yet so familiar. The room was very hot. Isabella was drowsy, half-asleep. She must go and sell her matches, or she would be beaten again . . . a cry of terror escaped her lips: *Sit there till I come again. Sit there. Sit there . . . .*

But struggle as she might, she could not move.

He will never come, she heard herself protesting. *He will never come, Hans. What am I doing here, after all these years?*

The portraits pinned upon the walls are faded and yellow now, and curling at the edges. Here and there among them a name, a signature, may be deciphered with difficulty — *Jenny Lind, Adelina Patti, Etelka Gerster, Isabella Weber* . . .

The aroma of hot coffee rouses her. She opens her eyes. She feels much calmer now.

'Is Madame recovered?'

'Quite recovered, Hans. I have had a long day.' Her eyes seek the curling portraits. 'Do you remember those days, Hans?'

'Very well indeed, Madame. Is Madame warmer now?'

'Much warmer, Hans. I almost envy you your comfortable little office. So different from the draughty stage out there.'

'If Madame is comfortable, it is enough.'

'Much warmer, and very comfortable, Hans, thank you. I could stay here for the rest of the evening.'

Hans gives her a quick glance.

'Madame is quite well again?'

'Indeed, Hans. I was cold.'

'It is not my business. It is an honour that you should condescend to visit me in my little office.'

'Honour, Hans, honour. Are we not friends? No honour is intended. I was Isabella when I first came into this room. I am Isabella still.'

Hans makes a deprecatory gesture.

'That cannot be so. Madame is *Prima Donna Assoluta*,' he protests.

'If that is so, Hans, it is due entirely to the goodness of nature and the generosity of friends, as you very well know. It is certainly no reason for a display of false pride, for conceit.'

Hans shakes his shaggy head.

'Pride? No, false or true. Conceit, no. But due respect must be paid to Art.'

Isabella smiles.

'Art is an artificial thing. It smells of the studio, the practice-room. It is divorced from Life. It is secondhand.'

'One must, needs, recognize achievement, Madame. And pride in one's achievement — legitimate pride — is sometimes a good and salutary thing. We do what we are called upon to do, to the best of our ability, and if by chance we do it a little better than others do, where's the harm in a little pride?'

'As one grows older one prefers a little humility, Hans. In any case, the definition of achievement is a matter of opinion.'

He throws a dark glance at the curling portraits.

'There are those who had no doubts about the matter, Madame.'

Isabella smiles naughtily.

'There are still one or two such, thank heaven.'

When the purport of her remark strikes her, she laughs aloud.

'There you are, Hans. You have moved me to a sort of conceit at last. Who am I, that I should criticize the opinions of others? Especially of my fellow artistes?'

'At least, Madame could never be ungenerous.'

'You flatter me, Hans.'

'I regret that Madame should think so. It is scarcely my habit to flatter. There are others who do it with more *addresse*.'

'That is so,' she agrees with some bitterness. 'You must forgive me, Hans. You yourself have ever been the sincerest of friends.'

Hans does not reply. He studies her face for a moment or two before handing her a cup of coffee. Something in her voice, in her attitude even, defies analysis, but causes him uneasiness.

'Forgive me, Madame,' he begins, and after a pause, 'if I say what I perhaps should not say.'

Isabella inclines her head, and waits.

'Madame is not happy about the part she is to sing tonight?'

'I chose to sing it, Hans.'

'I am aware of that, Madame. Indeed, the whole city is by now aware that it was your own choice. But the city does not know why.'

'And you do, Hans?'

She faces him directly. 'You are very percipient . . . .'

Hans does not reply.

'And you are very understanding.'

He continues merely to look at her.

'It is twenty years ago since you sang in your native city, and then you disappeared almost without trace. You sang *Alma Perdita*. That photograph is my souvenir, Madam, of that most happy time. For twenty years you have been lost to us. Now, without warning, and on the accession of His Highness, you accept an invitation to sing again . . . .'

'*Alma Perdita*,' she murmurs.

He observes that she has closed her eyes again, tightly, and that the knuckles of the hand which rests on the arm of the chair are white from the tension of her grip. When she opens her eyes again they are dark with pain.

'I said you could never be ungenerous, Madame, and but for the reason that you have never been other than generous to me, I would not have spoken as I did. It is after all no business of mine. My heart betrayed my head. I beg of you to forgive me in your turn.'

'There is nothing to forgive, Hans. Your opinion of me, whatever it may be, is justified by your knowledge of me. But, perhaps fortunately, you do not know everything.'

'That is possible, Madame; I am not curious.'

'Hans, I wished to hurt as I was hurt.' Her voice breaks. 'Now I think I am ashamed of myself.'

She puts her empty cup on the table, and with the necessity of simple control required by that movement she regains control of herself.

'It is too late to hurt, Madame, and of no avail. Now, it seems, you can only sing. Perhaps that will hurt the more. That is perhaps what Madame had in mind.'

'How wise you are, Hans. Would that I were as wise.'

'To sing is the best.'

'I can only sing,' she agrees. 'I shall sing tonight, and then perhaps no more. It is like to be my swan song. I am very tired, Hans.'

'Madame. Isabella. Do not say that.' Hans protests in horror. 'Do not say that within the theatre, or anywhere else, for that matter.'

Isabella essays a smile.

'Why, I do believe you are superstitious.'

'Who in the theatre is not, Madame? And many a true word is spoken in jest.'

'I was not speaking in jest. After this evening I shall sing no more. My mind is quite made up. You may be assured I have my reasons.'

'It will be a loss greater than the loss of kings, Madame.'

Isabella sighs. 'You are still flattering me, Hans.'

Hans shrugs his shoulders, and turns to the window opening on to the corridor. Along the corridors go the singers and the players. Each has a word of greeting for him, but, contrary to his usual habit, he does not reply.

'Hans has a headache,' they giggle as they pass on. 'Poor Hans.'

The fingers of the clock on the wall indicate a quarter to the hour, the time for her to leave the warmth and homeliness of Hans's little office and retire to her dressing room, where Maria is waiting to put her into that wonderful gown which caused almost as great a furore twenty years ago as her singing. So much was it admired, envied, coveted, that before the season was over it had initiated that famous series of Meissen figurines representing Isabella Weber in the character of Cordelia. So successful was this model that others followed in quick succession — Isabella Weber as *Gilda*, Isabella Weber as *Leonore*, Isabella Weber as *Amina*. They may still be found, in reduced numbers and enhanced prices.

A knock at the half-glass door calls attention to a diminutive messenger boy peering round a bouquet of roses as big as himself. The dark-red velvet of their petals is jewelled with crystals of snow.

'Personal,' shrills this small apparition, 'for Madame Isabella Weber. Have I permission to carry them to her dressing room?'

'Madame Weber is here,' Hans tells him. 'Give them to me. Such foolishness, in the middle of winter.'

Isabella half rises from her chair, her hand to her breast.

The boy puts his head cheekily round the door, and gazes open-eyed at his idol.

'Away with you,' Hans pretends anger, but gives the boy a coin as he takes the bouquet and carries it to her with all the gentleness with which old men handle flowers. Isabella drinks in their perfume.

'Are they not lovely, Hans? There is no note.'

Hans sniffs, with an expression of displeasure.

'There is only one place they could have come from, at this season of the year,' he remarks.

'But are they not lovely? Are they not perfect?'

'An unnecessary gesture, Madame. It is useless to fan a dead fire, if Madame will allow me to say so.'

Isabella continues to savour the perfume of the cool blooms. She reflects that though the fire within herself seems dead enough now, the flames had once burned fiercely and brightly enough. For her part, she has not striven either to keep them alive or to quench them. She has accepted the inevitable, since there was nothing else she could do. Nevertheless, at the back of her mind, there has always been the niggling thought that someday, in some way, she would repay.

It is indeed useless to fan a dead fire.

*

A bell rang loudly, startling her.

Hans smiled, and reassured her.

'It is the signal,' he told her. 'The orchestra is now taking its place. In a minute it will play. The curtain will go up, and the opera will have begun.'

She had really no idea of what he meant by this laconic summary. She knew, beyond doubt, that her Englishman was somehow involved, making music with his violin so that Jenny Lind might sing. In her ignorance the child imagined that opera was Jenny Lind, and that Jenny Lind was opera.

The sound of music came faintly from far away.

'How wonderful it must be,' she said.

'The only people to whom it can be wonderful are those who pay to listen to it,' said Hans, 'and not always to them. For those who take part, it is rarely anything but hard work, year after year of struggling to achieve perfection. They are only the singers. Then comes the composer; in tears and perspiration, scratchings out and rewriting, he finishes his score, and perhaps manages to interest someone in it. The musicians play it, the singers sing it, well or more frequently, let us admit, indifferently. As for the rest, the scene shifters, the lighting men, the dressers, the cleaners, they all give their hearts to their work. But the audience . . . . What does the audience do? The audience criticizes. The audience is disappointed that it was not done this way or that. The audience does not like the dresses, the scenery, the soprano's high notes, the bass's low notes, the conductor's interpretation. No, the audience could do it better . . . .'

The child was not listening to him. Her outward eyes were blinded by an inner vision of music and splendour.

'Come', said Hans, on a sudden decision. 'You shall see what is it like for yourself. Put your matches on the table.'

She followed him along several corridors until they came to a big notice printed in red; it forbade them to pass any further. Hans drew the child into the shadows, placing her where she could see all that was going on.

'Do not move, child,' he whispered fiercely. 'Do not move at all.'

The injunction was scarcely necessary. She listened spellbound to the glorious music; gazed entranced at the majestic scenery. On the stage a man and a woman were singing together, and while they sang a great light shone about them.

When they stopped singing, and stood quite still in the middle of the light, from somewhere out of her range of vision there came a thunderous noise of clapping and voices calling. The two singers released themselves from their embrace and bowed over and over again. Then suddenly other people ran out of the shadows all about her and began to take the wonderful picture to pieces. Lights flickered, became white and blue and green. Dust rose in little clouds to be wafted away on vagrant currents of air. Mountains rose and moved majestically and fearfully. Deep and gloomy caverns appeared. The clapping continued.

The two singers came towards her, passed her, not seeing her. As they passed her she realised, with a momentary feeling of disappointment, that they were ordinary people; they spoke to each other, and laughed a little, and disappeared.

The magic, however, did not disappear; this was the 'opera' she had heard people talk about. This was the wonderful thing the beautifully dressed people who sometimes bought her matches came to see night after night.

Almost sobbing with excitement she ran back along the corridors, seeking the warm safety of the doorkeeper's room.

'It was wonderful,' she exclaimed. 'Such beautiful music; such beautiful singing.'

'Wagner,' said Hans, in a tone of voice that did not express complete pleasure. 'You have heard of Richard Wagner?'

The child shook her head.

'A-ha. Such goings-on, that's what I say. A lot of noise, Wagner is. Too much wind, one way or another. Mozart, now, is different. More gentle, more refined. Delicate you might say.'

This elementary exposition and criticism of the composer's art was quite above the child. Her head was still sounding with the music she had heard. Noise or not she did not know. If it was indeed noise, it was the most beautiful noise she had ever heard, quite unlike other noises she recognised, the singing of birds, the rippling of water in the fountains, the sound of the wind. These certainly gave pleasure, and were beautiful, but the music of humans made one wish to laugh and cry at the same time.

She found this idea impossible to explain to Hans.

'There is music and music, child,' he replied. 'Some you like, some you don't like.'

The child paid no attention to this. She was wrapt, in her own thoughts. For her, from this time on, music was music; at its best or at its worst, should there be such a distinction, it was the expression of a soul blessed or tormented with a more than mortal vision.

'Hans,' she said with sudden decision, 'I shall sing opera.'

'There, now,' Hans replied. 'We shall see.'

She was now pale, and tired. He gave her more coffee, and added a little cognac, this being in his understanding a specific remedy for all ills. It brought a temporary sparkle to her eyes; but suddenly she fell asleep.

Herr Smith shook her gently.

'Isabella,' he said. 'We must go now. It is all over, child.'

'I have a coat here,' said Hans. 'It is very cold outside. Good night, little Isabella.'

Decision still glinting in his ill-fitting spectacles in spite of the long evening, Herr Smith and the child went out into the freezing air, into the darkest, most twisted part of the old city. He had at this time no clear idea of what he was about except, as he continually muttered to himself, something ought to be done. He was in fact, following an unanalysed inner compulsion. But even as he walked with her through the dark night he told himself that he was a quixotic fool; that no good could come of it; that his English sentimentality had been betrayed by an attractive immature voice and by one more heartbreaking piece of evidence among many of ill treatment among children. It pained him to see how wearily she now shuffled her feet along, calloused, dirt-engrained feet in a pair of broken boots tied round with string because they were too big. She moved as though moving were a supreme effort; her steps became more laggard as they went. Her thin body was bent, her shoulders hunched. From time to time she was shaken by a cough.

With one hand she clutched the unsold bundles of matches in the tattered skirt of her dress. With the other she clasped his hand, not tightly, but firmly with a dog-like trust. He volunteered no information, and she asked no questions. What he was thinking about she had no idea. She led him without hesitation through streets which turned and twisted and grew narrower and darker; beneath flickering lamps suspended on ropes at corners; under arches where gibbering shadows came ominously together and whispered and melted away as they approached; past sudden openings through which the lighted windows of the castle on its hill were momentarily to be glimpsed hanging above them in the cold frozen sky.

At one of the most disreputable warrens she led him up the malodorous wooden stairs. On the first landing, where the light from the lamp hanging over the portal below no longer helped his stumbling steps, she struck a match so that her companion should not become entangled with the piles of rubbish piled there. On the fourth landing she extinguished what was left of the match. She opened the door immediately before her, and went fearfully in. Not knowing quite what to expect, he followed her in.

The room was no more than an evil-smelling closet with a high window. In one corner there was a tumbled bed; in another a heap of rags with some pretence to being a sleeping place. In the middle of the room a wooden table had as companions two unstable chairs, and beside it was a charcoal brazier. In the draught from the open door the flame of a candle in its stand flickered and jumped, and a square of paper pasted over a broken windowpane rattled as if a skeleton were cracking its ten bony fingers.

One of the two bundles crouched at the brazier moved.

'Close the door, child. Do you wish to freeze us out altogether?'

The child closed the door quickly, but did not advance further into the room. Herr Smith, who was still holding her by the hand, could feel her trembling.

The other figure by the brazier now rose to its feet, and began to fumble with the belt about its waist.

'How many matches has little Bella sold today?' it said with a belch. 'Come, child. Tell me, how many matches have you sold today? And where is the money? Give me the money.'

The child shrank from him, trembling.

'And who is this?' he shouted. 'Do you bring strangers into your home?'

He swayed before them, his breath heavy with the fumes of liquor.

'Answer me. Who is this stranger, and what does he want?'

He lurched towards the child, and halted and swayed but a finger's breadth from Herr Smith. He raised the belt, but, catching Herr Smith's eye, thought better of the gesture, and let it fall by his side. The old woman turned in her creaking chair. She stared at Herr Smith and the child with cunning eyes. She said nothing.

The candle flickered violently in a draught, throwing menacing shadows across the walls and the discoloured ceiling. With every breath the man exhaled a new and distasteful odour was added to the foetid atmosphere.

'What do you want, stranger?'

'This is your child?' Herr Smith pulled her closer to him as the man made a threatening move.

There was no reply. They stared at him as if the question had taken them unprepared.

'This is your child?' Herr Smith repeated, raising his voice.

The woman eased her bones and answered him.

'As you say. She is our child.'

The man turned upon her.

'Be careful what you say, woman,' he shouted. 'Our business is our business, and of no interest to anyone.'

He turned upon the Englishman and the child again.

'Why do you ask? What is it to you whose child she is?'

'I am interested, that is all, in spite of the fact that it is none of my business.'

'Who is this you bring home so late?' The fellow roared at the child again. 'By what right does he ask impertinent questions? Answer me.'

'He . . . he is an Englishman,' the child faltered. 'I do not know his name . . . .'

'An English, eh?'

'He plays music at the opera.'

She glanced beseechingly at Herr Smith. He pressed her hand reassuringly.

'So. And what does this English want?'

'I do not know, father.'

'Bah. You know nothing. You are a simpleton.'

'I want the child,' said Herr Smith, suddenly.

Silence fell oppressively on the room, a silence in which the splitting of a piece of charcoal in the brazier sounded like the crack of a pistol. The eyes of the two old crones glittered at each other across the smoking candle flame.

'So . . .' said the man at last, and his tone was unpleasant, suggestive.

'I want the child,' repeated the Englishman, battling with an uncertainty which had begun to weaken him. 'I wish to . . . to adopt her.'

He pronounced the word as though it were unfamiliar.

'Adopt?'

The old woman shrieked the word in echo, giving it at the same time a kind of obscene emphasis. 'Adopt, our little Bella?'

Again the two stared at one another. Their eyes narrowed; their yellow faces cracked into a thousand pieces; harsh laughter cackled in two throats long unaccustomed to laughter. They laughed until the tears ran down the old woman's cheeks.

Between his paroxysms of laughter, the man roared to his wife: 'He wishes to adopt her. Do you hear that, Mama? This English, this fine gentleman, wishes to adopt our little Isabella, the pride and joy of our old age.'

Again he roared with laughter.

Suddenly he stopped laughing.

'No!' He flapped his hand loudly on the table, almost dislodging the candle from its stand.

'No!' The woman echoed shrilly. 'She is our child. Why should you want her?'

Shrill though her voice was, it seemed to Herr Smith that it lacked conviction. She looked at her man. Together they talked in an incomprehensible dialect, and their heads nodded in unison, like the heads of clockwork toys.

The man turned to Herr Smith. The tone of his voice, when he spoke, was more moderate.

'What is it you know, Herr English?' he said, almost unctuously.

Herr Smith was taken aback. Clearly there was something he did not know. Clearly they were disturbed and worried.

'What do you mean?'

'About the child. Has someone talked?'

'Talked of what?'

They stared at him in silence. The man belched loudly.

'I am sorry for her,' said Herr Smith. 'She was cold, tired, and hungry. I could take better care of her.'

The old woman shrieked at him again.

'We have taken care of her, within our means. We are poor peasants, Herr English. We do what we can. She has a roof and a bed. She shares our food . . . .'

'And you beat her,' put in Herr Smith.

They stared at him again, in apprehension. A miasma of alarm, of fear, crept about the foetid room.

A feeling of depression overcame Herr Smith. He moistened his lips, salty with disgust at the situation in which he found himself. For several heavy minutes nothing was said on either side. But, by the expressions on their faces it was clear that something was stirring in their minds.

He waited for them. They waited for him. Between them the child trembled, fearful lest the promise of happiness held out by Herr Smith's hitherto puzzling behaviour should be nullified by a drunken perversity.

'A roof beneath the tiles. A broken window,' he said at last. 'A bed that is merely a bundle of rags on the boards. And as for food . . . .'

He made a gesture of disgust.

'I will buy her from you,' he told them. 'I will give you money for her.'

Their eyes brightened. He had decided that this was a move they were expecting; he could not think what madness impelled him to make such an offer. He could only feel the ridiculousness of the situation. It had not even the dignity or the necessity of the Roman slave market.

The relief that now showed in their faces was evident. It was quite clear also that they thought him mad, for to them no reason was apparent for anyone to be anxious to get possession of a female child who was ugly, who had no pretty ways, who had no more flesh on her bones than a wild rabbit; who could neither hew wood nor carry water; who was useless even at selling matches.

Certainly this Englishman must be mad.

The man swung his belt to and fro from force of habit, and cracked it between his great hands.

'I will buy her,' repeated Herr Smith, watching these threatening motions. He felt his courage in making the offer weakening.

The man grinned, showing his deformed yellow teeth. He took up a bottle from the table and drank deeply from it.

'So,' he said, wiping his mouth with the back of his hand. 'So, you will buy her, our little Isabella. Very well, Englishman, your price? Name it. How much are you willing to pay to take from us the pride of our hearts, the support of our declining years? How much money will recompense us for the loss of our little loved one, in your opinion?'

He leered at the Englishman, and belched again.

Taken by surprise at this sudden change of attitude, Herr Smith's heart tightened in cold anger.

'There is no pride in her here,' he said in contempt. 'There is no love, no compassion. The child lives perpetually in fear.'

The man shouted him down brutally. The relief in his manner was evident.

'You talk too much, English. We waste time. Your price?'

The Englishman hesitated, fully aware of his meagre means, and the game would be lost or won on his ability to estimate accurately the extent of their cupidity. Even now, he thought, it was not too late to change his mind, to deny his foolish sensibility and to substitute a little sense. The child was nothing to him.

'Come, English, make a price. Twenty Thaler, shall we say?'

The woman screamed in anger.

'Twenty Thaler. Oh, but you are quite mad. The price of a goat.'

The bully roared again with laughter, and took another draught from the bottle. The wine ran in rivulets down his unshaven chin.

'*Gott in Himmel*. You are right, woman. The price of a goat. It is heavy, that. But a goat is better. A goat gives milk. What can she give, eh? Tell me that.'

He choked on the wine between his laughter, and threw the bottle violently into a corner.

'What can she give, eh, Master Musician? Look at her. Figure to yourself what satisfaction you will get from her.'

The Englishman looked his distaste at the other's coarseness.

'Twenty Thaler, I say. Come, show us your gold.'

'Forty Thaler, fool,' the woman screamed.

The child's eyes were wide with fear. She shrank back against the filthy wall while they fought and scratched over her price.

'Forty Thaler. She is not worth it. I would not rob any man.'

'Your child,' the woman reminded him.

'So you say, woman. But what matter. She is quite useless, and a burden. She sells no matches; she does no work of any kind; her head is full only of what she calls singing, that is all. She can think of nothing else. When she is in a good mood she caterwauls all day. How annoying it is. By Heaven!' he swore. 'How bored I am with all this. Twenty Thaler, Master Musician, and the brat is yours.'

The woman tore at her hair in anger.

'Twenty Thaler, he says. Fool, drunken, besotted. Twenty Thaler. *Du lieber Gott.*'

He grinned at her, and belched again.

The woman grabbed the Englishman's hand in her grimy paws. 'My man is sick,' she whined. 'He does not know what he says. We are poor, Herr English. We are excessively poor. The child eats more than she earns. It would certainly be better if you took her.'

Her tone changed. Her eyes were greedy, calculating.

'But not for twenty Thaler. No, forty Thaler would be better.'

She tightened her grip on his hand.

'Forty Thaler, eh? Forty golden Thaler, Herr English, as recompense for our love and care.'

The man was drinking again from a fresh bottle, leering and grinning as they argued. Herr Smith thrust the woman from him. She dragged the trembling child forward into such full light as the flickering candle provided.

'Look at her,' she cajoled him. 'See this hair, it might have been spun from molten copper. Look at those eyes. The grass is not greener. See these limbs . . . .'

Shamelessly she lifted the tattered hem of the child's skirt to display what she intended him to perceive was a promise of beauty in the thin legs beneath.

'Thirty-five Thaler, Herr Englishman. As God lives, you will not regret it. May man is a fool. I know what he does not.'

The man roared with laughter.

'The little food she eats will not cost you much, nor a little red wine now and then. We are indeed poor. My husband is a good-for-nothing. For you, thirty-five Thaler is nothing. For me, it is riches untold. Thirty-five Thaler, Herr English, and take her.'

Her voice fell. She almost whispered. 'There is the best blood of Ruritania in her.'

He ignored this statement, which he considered to have been made simply to extract the greatest possible sum of money from him. Wishing to end as quickly as might be a situation which had become nauseatingly intolerable, the Englishman took out his purse. The woman's eyes brightened at the tinkle of coins. She clutched his elbow, digging her talons deep into his flesh through the folds of his coat. Her head nodded in rhythm as he counted out the pieces on to the table.

'Twenty-seven, twenty-eight, twenty-nine . . . .'

'Thirty-five, Herr English . . . thirty-five.'

. . . 'Thirty.'

She tightened her grasp on his arm.

'Thirty-one . . . thirty-one, thirty-one . . .' She was almost voiceless from frustration.

He put his purse away. 'Thirty Thaler is all I have. Take it or leave it,' he said with finality.

For a moment it seemed that she was about to loose a tirade of abuse at him. Then she sighed and began to pick up the coins one by one, telling them over with soundlessly moving lips.

When the last coin was taken up, Herr Smith turned to the door.

'Wait,' cried the woman.

From a box in a cupboard she took out a purse, into which she put the thirty Thaler. From the same box she took, glancing furtively at her husband a piece of paper, brittle and brown with age. This she pushed into Herr Smith's hand, closing his fingers over it.

'Guard it well,' she said. 'It is of the greatest importance.'

He put it in his pocket, saying nothing, but feeling a sickness in his heart. He took the child's hand once more, and together they stumbled down the stairs into the cold darkness of the late winter night.

The darkness was illuminated only by the lights in the windows of the distant guardroom of the castle and by the pinpoints of the even more distant stars. The guardroom lights were still and permanent, but the stars moved in their courses, weaving about the heavens endless spirals of light, just as the great bell of St. Michael wove a spiral of sound about the Square, the sleeping houses, the silent streets, and trembled into silence.

*

As the last stroke vibrates to its end, the percussion player sits back with a satisfied air on his stool, and folds his arms. He has quite a long rest before him. He hears without hearing the echo of that last tragic line: *Come la vita e sempre vuota . . . .*

He permits his eyes to wander for a short time over as much of the darkened house as he can actually see.

Beyond the edge of the curtained orchestra rail the vast honeycomb stretches away and upwards, filled to capacity with intent, faceless worshippers, one would like to think, at the altar of music. The pillars are twined about with out-of-season flowers. The boxes are draped with white and purple. In the Royal Box, the epicentre as it were of this great affair, sits His Majesty, the newly-crowned Karl Ernst Maria Frederik, King of Ruritania. His face is white and grim, unsmiling. He smiles rarely these days. According to 'reliable sources', he has not smiled for twenty years. Even at his wedding, so long ago, which should have surely been a happy occasion, one supposes, it is said, his face was the face of a man in a trance.

The townsfolk were quite unable to explain it.

'He was once a happy young man,' they said. 'He had his hunting, his banquets, his parties . . . Now look at him. A most unprepossessing young man our prince has turned out to be. The joy of life no longer flows in his veins.'

The percussion player cannot explain it either. Being of a somewhat Socialist disposition, he is not really concerned with the personal troubles of Royalty. He supposes that if one were burdened with the family history of the Elphbergs; if one were forever disciplined in the harshest way by a tyrant of a father; if one were ordered to marry that mare of a Margravine simply for dynastic reasons; and if one were to all intents and purposes immured in that castle up there on the rock, one would find very little to smile about.

The percussion player shrugs his shoulders, and prepares for his next entry, his eyes on the *dirigeant*.

*

With the same unchanging face Karl Ernst Maria Frederik has borne the weight of increasing years and duties; the opposing burdens of an ailing intolerant father and a barren shrewish wife. Today he has sat, with the same joyless demeanour, through the long hours of the Coronation ceremony. Now in a situation over which he has little control, he sits in the Royal Box at the Opera, bitterly savouring the fact that Fate seems to present sometimes punishments worse than death. His Consort is not aware of the trend of his thoughts, but it would make little difference if she were. Vapidly she conveys a success of bonbons to her mouth and chews automatically. She waves her fan languidly to and fro.

Behind her, in the perfumed dimness of the Box, her attendant ladies exchange significant glances and smile provocative smiles at His Majesty's Gentlemen-in-Waiting.

But if His Majesty's face is as grim and unsmiling as ever, his eyes are far from lifeless. They move in his still face with every movement of that figure on stage, so diminished by distance, yet able to fill every corner of the House with glorious sound. They are the eyes of a soul suddenly released from bondage, a soul renewed by a vision which had become increasingly tenuous with the passing years. Years are long, and memory fallible. It was not to be expected that Karl Ernst should remember in every detail the perfection with which

Isabella Weber, a young singer unknown in her professional capacity in Strelsau, astonished and took by storm a city whose standards of musical appreciation were proverbial — at least among its own citizens. *When Strelsau applauds, artistes weep with joy* was the cliché they had invented for themselves. In common with his subjects, His Majesty had been overwhelmed by the consummate artistry of the acting and the perfection of that well-remembered voice.

How can he not remember, when that first clear note pierces his heart like the point of a knife?

*

Signor Andelli raises an expressive eyebrow, and flicks his baton.

The timpanist executes a delicate roll on his drum, *pianissimo*, before the orchestra introduces a reprise of the humming theme to which the village worshippers have entered the church. They are now leaving it, and the music rises and falls, hauntingly beautiful, conveying a feeling of peace and tranquillity. Suddenly the mood changes; the music is charged with agitation and fear. The villagers gather about Isabella, a stranger to them. They are curious, questioning. Who is this stranger in distress? And why is she in such distress?

'*Chi e?*' they murmur among themselves. '*Ch'e succeso* — what has happened?'

'*Help me*,' sings Marella in a small frightened voice, 'my lady is ill. Is there no one will help me?'

The violins soar on wings of hope.

Father Abbot stoops to lift Cordelia to her feet.

'*Mia figlia,*'he sings in a sonorous bass. '*Dici qui es . . . .*'

Supported by his encircling arm, she endeavours to concentrate on his kindly face, to know it, lest it should disappear into the darkness for ever. Her eyes seek his eyes, seeking to draw some spiritual succour from their steadfast kindliness.

'*Dici qui es,*' he sings again. '*Poverina mia . . . .*'

Cordelia tries to smile through the haze in which she is caught up. She must answer before the darkness descends upon her completely. But when she answers her voice has lost all feeling, has been drained of all colour.

'*Mi chiama Cordelia* . . . Isabella . . . .'

Wildly she looks about her, fearful of the thrusting anonymous faces on all sides.

They murmur. They mutter:

'Cordelia? *Qui e Cordelia?*'

The abbot holds her gently, paternally. Sick with pain she clutches him tightly. The world, its light and darkness, revolves uncontrollably about her. The anonymous faces advance and retire, leering, threatening.

'How come you are here, my daughter?'

She stared at him for a moment scarcely understanding the question. Then with a sudden movement buried her face in the fold of her cloak, grasping his hand so tightly that he almost cried out.

'My name is Isabella.'

Confronted by the chilling animosity of the onlookers and the near-hysteria of the child, Herr Smith verged on the hysterical himself. He was at a loss to account for the child's outburst.

'Isabella,' he said gently, 'if you can walk a little way, we will find some nourishment for you.'

Still holding his hand, she allowed herself to be taken to Hans's little room.

*

The memory of an unhappy childhood in this city where the very stones of the streets are still familiar has remained bitterly sharp throughout the years. How strange it is that it may be visited now by invitation, an honoured guest. And in between, what?

Most importantly, she remembers, she had regular meals of plain but substantial food, and a proper bed to sleep in, in a comfortable room. She also had new clothes, practical rather than decorative; but to a child who had previously too little food and too few clothes, it was paradise enough. She no longer had to sell matches, and she was no longer beaten for the smallest fault. Nevertheless, there appeared by degrees a gap to be filled. Warm, well fed, well clothed, and properly rested, she had nothing to do. Relieved of the necessity to go out into the streets, and because she was left alone for the greater part of the morning and evening when there were rehearsals and performances at the Opera, the child was left to her own resources. For want of anything better, she began to rearrange the contents of the meagre apartment. This, she discovered, gave her some feeling of satisfaction. After rearranging the furniture, she polished it, over and over again. It had never had such a surface on it. Polishing became a weekly routine. Then there were the two windows to wash. From the balcony which joined the windows she could see the tower of St. Michael above the roofs of the intervening houses, and hear its great bell. The bell was indeed her only timepiece. When it sounded the hour, she began her self-appointed task; when it sounded in like manner a second time, that task was finished, and she began another.

It was quite some time before Herr Smith, wrapped in his almost permanent abstraction, realised what was going on. He lived in a world in which music and his part in it were the mainspring of existence. It only came to him gradually that his establishment was not what it used to be; that it had begun to resemble the spick condition it had had when his wife was alive. He felt near to tears as he praised the child for her application; she renewed her efforts thereafter, and in his turn he discovered once more the joy of returning to his hearth after an absence short or long. He praised the child for the cleanliness and neatness of his rooms and the delicacy of his meagre table; he complimented her, somewhat awkwardly, on the care she bestowed on her toilet and her dresses. Her hands had become quite clean and fine.

He was touched by all these things more than he could have said, and he sought for little ways in which he could show his appreciation. He bought her packets of bonbons, and sometimes a special cake, a length of ribbon, bits of lace, a box of coloured threads, cottons and needles. His totally abstract mind rarely rose to greater heights of inspiration than was demonstrated in these gifts, although on one occasion he triumphantly carried home a doll. He felt more than a little foolish when he gave it to her, uncertain as to how she would receive it. It was not a particularly beautiful doll, and he was quite surprised when she declared it was the most beautiful thing she had ever seen. She wept over it a little. It became her constant companion. Throughout the day she talked to it, alternately scolding and praising its silence and its garrulity; its cleverness and its stupidity. At night it slept, unblinking, in her arms.

It rejoiced in the name of Araminta.

'Araminta says this. Araminta says that,' she would tell him gravely as they sat at supper before he went off to the Opera. 'Araminta is really a very intelligent person, you know.'

Araminta was a third ever-present member of the family.

To his great relief the child was no longer so timid or so silent. With some hesitation at first, and then more frequently, she called him 'Papa', and he felt a ridiculous lightening of the heart. She relished her food; her limbs were no longer so painfully thin, and she no longer coughed so frequently. She became, in his eyes, almost pretty.

But she no longer sang.

Herr Smith was extremely puzzled about the situation. All too well he recalled the time when he would hear her voice, singing the songs of the people, and he would go about his business with vague longings to do something about it. 'She sings like an angel, Hans,' he had said. Now she no longer sang. His Quixotic venture had been based on the fact that the child sang like an angel. Now, she did not sing.

'She has a good home,' he told himself, 'she has good food, she has clothes enough and to spare. Why then does she not sing?'

His nose twitched in bewilderment, and the wrinkles deepened on his forehead.

'Why did I do this thing?' he asked himself many times. 'Why did I burden myself with this child?'

And immediately he was ashamed of himself. 'One takes pity on small helpless things,' he excused himself angrily.

It seemed also to the child that he had, in some way, lost the original interest that he had shown in her. Sometimes she felt, although she was no longer badly treated, that her presence had no meaning for him. She was at a loss to know what to do or say.

It was not until many years later that she was able to appraise fully his part in her life, and by then it was too late. If she had understood, if she had been able to understand, what effort of will it had cost him that far-off evening to face who knew what physical attack, or at best, insults of the worst kind, or if she had known that behind the seeming disinterestedness he was making and rejecting plan after plan for the furtherance of her welfare, she might not have judged him so harshly.

The solution to the problem was, in its own time and its own way, simple. She was hungry for music, although she could not have said so. He went every evening to the Opera, carried his violin, that extraordinary extension of his right arm; why could she not go to the Opera?

She was so surprised by her own request as he was to receive it, and would have withdrawn it. There were difficulties. He could not afford to pay for her; and in any case it was unthinkable for a child to attend the Opera alone . . . .

It was Hans who came to the rescue. It was Hans who secreted her first in his little office by the fire, as he had done before. It was Hans who stationed her, against all decree to the contrary, in a dark corner in the wings with the repeated injunction '*not to move*'. It was Hans who gave her coffee once more in the intermission.

It was Hans who jokingly asked if she had yet learned to sing, and was surprised by the vehemence of her reply.

'I shall sing, Hans.'

When this adventure was over, and Herr Smith and the child were on their way home, he listened with some interest to her account of the new world once more opened to her through the narrow door. He thought little of her childish enthusiasm at the time, but when, the following day, she intoned almost correctly two of the evening's arias, he was quite unprepared for the demonstration. Thoughtfully he took out his violin and requested her to sing them again to his accompaniment. Carried away by mutual enthusiasm they went at it for an hour of repetition, fiercer and more passionately, till an angry knocking from overhead warned them that it was time to stop.

He put aside his violin, and for the first time embraced her in a fatherly hug.

'My child,' he said. 'I am a stupid old man. I should have known. Can you ever forgive me?'

Not knowing what she had done that seemed to please him, nor for what reason she had to forgive him, she remained silent. Because he was pleased she too was pleased, and, as she understood it, loved; and because of this, she wept, clinging to him.

*

The music continues, a slow sad *pavane*, while the chorus groups and regroups about the two central figures on the stage. The Abbot supports the half-fainting Cordelia tenderly; his shadow enfolds her. He applies all the extemporizing art of the trained actor to the task of preventing the audience from realising that all is not quite as it should be. Nor can he allow his own bewilderment to be seen. Never, so long as he can remember, has any artiste given way to personal emotion on stage.

He can say nothing. The lightest whisper would be magnified tenfold about the auditorium. Nevertheless, the intensity of his gaze upon her reflects his silent anxiety and apprehension.

Cordelia raises her face to his. She smiles, weakly, held in his strong arms. With the greatest economy of movement she inclines her head, indicating that all will be well. Her eyes are still wet with tears.

The moment of vertigo passes; he puts indecision from him.

'*Dice qui e . . . .*'

'*Mi chiama Cordelia.*'

The voice is low, but it can be heard clearly on all sides.

A movement of relief passes through the audience. 'What an artiste,' they think to themselves. 'Superb.'

On a single note from the bells, the music begins to swell in confidence. Movement possesses the company on stage. Only Cordelia, still histrionically — as they think — weak and from the rigours of her fictitious wanderings, remains motionless, supported by the fatherly abbot. The music swells, spreads, weaves back and forth, modulating into a contrapuntal duo for basso and soprano. The two sing in question and answer; the music continues in statement and contrastatement, filling the House with golden sound.

He questions, she replies; he pleads, she rejects his pleading. He advises. She breaks away from him angrily; she is alone, friendless. She will go her way, as she came.

'*Senz' amore, senza vita . . . .*'

In the sea of faces before her not a muscle moves, save in one. Her Majesty Amelia Viktoria Konstanzia Camilla petulantly chews her bonbon. Petulantly she agitates the air about her with her fan. She is excessively bored, and is privately of the opinion that all this posturing, strutting and declaiming is the sheerest nonsense. And as for music, a pretty waltz is much pleasanter and far more interesting. She would not have consented to assist at such pretentious mummery as this had it not been for the fact that she wished to see with her own eyes this paragon of the Opera, Madame Isabella Weber.

Now she has seen, and heard. Her malevolent eyes look sideways at her husband and king. Allowing for the cares of office and the weight of years she finds him still handsome enough (taking into consideration the Elphberg nose prominent in all the family portraits from Black Michael onwards) but thicker set than he used to be. How handsome he was then. How many hearts must have beat faster at a word or smile from him. Even her own heart had seemed to swell dangerously on being introduced to him. 'Cousin,' he had called her. 'My pretty cousin.' She cannot even now remember whether he was being kind or patronising.

She has never been under any illusion concerning her handsome husband. But it has not been her business to have illusions. Her business has been to do her duty according to the desires of her man and the expectations of the State. If her tacit acknowledgement of such a liability has produced no positive result in twenty years she is entirely uninterested. She is sure he has had some consolation; certainly — she smiles a secret smile — she has had hers.

No one is any the worse.

Certain rumours have caused her some uneasiness, rumours involving in a strange way His Majesty Karl Ernst Maria Michael Frederik and that woman down there on the stage. Consider the situation as she might, she could arrive at no reason for this rumour. What has a vagabond opera singer to do with the King of Ruritania? That is a nut that would indeed bear cracking. The importance of that wretched singer in the life of the city has in the last few weeks been altogether disproportionate. Her Majesty's eyes grow narrower as she perceives that every movement *that woman* makes is followed by His Majesty's full and undivided attention.

It is not to be borne. With a movement indicative of frustration and anger Amelia Viktoria puts another bonbon in her mouth and chews upon it viciously.

Her ladies nudge one another, and suppress their giggles with difficulty as they observe the increasing agitation of the royal fan before the royal face. They are by virtue of long and sometimes painful experience adept at reading such thoughts as Amelia Viktoria may have, at interpreting such movements as her torpid body allows her to make.

In the limelight the duo becomes a trio. Lisa, in her role of Marella, Cordelia's attendant, demonstrates that she possesses a warm mezzo-soprano voice. She appears to be nervous, overshadowed as she is by Madame Weber and the bass Demetrio Apostolos.

The audience is intrigued by this girl, the only newcomer in an established cast of international repute, but it is to be seen how, as she sings supporting the Abbot's commanding bass against Cordelia's tragic uncertainty, she gains confidence and presence.

His Majesty sits straighter in his chair, his hands gripping its arms in nervous tension. In his mind there is the faint vibration of some chord of recognition. He observes that she is tall and straight, beautiful in a youthful way. Slimmer than Madame Weber, naturally, but with the same carriage and posture and, intriguingly, the same copper-gold hair.

Amelia Viktoria's sharp little eyes have noted that since the beginning of this trio her husband's face has assumed an expression she has never before seen. She stops chewing. She swallows. She raises an imperious forefinger.

A Lady-in-Waiting rises and curtseys.

Her Majesty points with the same finger directed at the distant Marella. The girl — and indeed all the protagonists assembled there, are almost invisible in the gloom of the approaching stage night, but her voice rings confidently and strongly in the trio.

'That girl,' Her Majesty says in a loud whisper. 'Who is she?'

Karl Ernst moves uneasily. There is a movement in the audience as if a wind had blown over a field of corn.

'Your Majesty,' whispers the Lady-in-Waiting. 'Nothing is known of her. It is understood that she is a protegée and pupil of Madame Weber, and tonight makes her debut. It is said that Madame Weber insisted . . . .'

'Madame Weber. Madame Weber. Upon what did Madame Weber insist? And why? Who is Madame Weber to insist or not to insist upon anything? There is altogether too much of Madame Weber.'

Her Majesty sports her fan in frustration.

'The name of the girl. Her real name?'

Amelia Viktoria could not have done better had she been down there on the stage herself. Faces turn in the dim light towards the Royal Box.

'It is understood, Your Highness, that she calls herself Lisa Weber,' the Lady-in-Waiting whispers.

'It is understood. Understood. Let me have less of these understandings. Her name, Fraulein . . . . If you please.'

'Her name, Highness, *is* Lisa Weber.'

'Ah,' breathes Amelia Viktoria.

'But it is understood . . . .'

A malevolent glance stops her short. Amelia Viktoria grinds her teeth and grins horribly.

'Fraulein von Plessy-Dortheim, I shall have you beaten.'

Fraulein von Plessy-Dortheim sketches a curtsey.

'Your Majesty.'

'Continue, Fraulein.'

'Madame Weber adopted the girl when she was but a few days old. It is said,' added Fraulein von Plessy-Dortheim, a shade defiantly.

'So. And where did this interesting event take place?'

'In England, it is . . . .'

'Fraulein. I perceive you are somewhat of an authority. What is the age of the girl, would you say?'

Fraulein considers.

'Perhaps twenty, twenty-one. It is difficult . . . .'

'That is enough, Fraulein. You may return to your place.'

Amelia Viktoria flaunts her fan again, but it is of little avail. The surrounding air grows hotter and heavier. It presses upon her, a warm sticky cloud. She feels a sense of depression which she cannot explain.

She turns to her husband.

'That is certainly of interest, is it not, my dear? The little girl with the copper-coloured hair, Lisa, little Lisa Weber, *adopted* by the great Isabella Weber. And she may be twenty, or twenty-one. What is your opinion on these matters, husband?'

His Majesty indicates the programme lying neglected on the plush. He is very obviously out of patience with his consort.

'I have no opinion. Her name is already known to me,' he points out shortly. 'As for the rest of the gossip . . . .'

Amelia Viktoria achieves another grin, baring her vulpine teeth. The expression sits ill upon her hot, rotund face, yet expresses undoubtedly her true nature. She considers at length the Elphberg features of her husband.

'Strange,' she murmurs through her smile. 'Strange that you have not heard that she was *adopted*, is it not?'

'I cannot imagine, Madame, that the private life of Madame Weber has any interest for anyone not within her private entourage.'

Amelia Viktoria chooses another bonbon.

'For my part,' she observes, 'I cannot understand why so handsome a creature as Madame Weber had need to adopt a daughter.'

Karl Ernst appears to pay no attention to this remark. His eyes are fixed on the distant figures. In the overwhelming glory of those voices — of *that* voice, the unlovely voice of Amelia Viktoria is like an icicle in the sunshine.

The trio ends in a magnificent declamation of persuasion, pleading and rejection.

Signor Andelli throws down his arms; the music stops.

In the unexpected silence — that pause which a generation of critics has praised as a dramatic ultimate worthy of Beethoven himself (but without the slightest hint of condescension) and condemned (without the slightest comprehension of its dramatic import) — the visual impact of Cordelia's last gesture of hopelessness on her last rising note is as pregnant with meaning as the musical phrase itself. On the occasion of the first presentation of *Alma Perdita*, the critic Hohenberg wrote in a fulsome article:

> 'There are musicians who use too many notes, and musicians who use too few. The composer Andelli, with an understanding which seems to be derived from St. Cecilia herself, uses precisely the right number. In addition, he knows precisely when to stop using them. This punctuation, this moment of silence before the next area (i.e. *Peccavi*) seems to sum up the whole of

Cordelia's life. It is also, it seems to me, the perfection of musical achievement. The orchestra stops on an unresolved diminished 7th (this alone has caused some raising of so-called musical eyebrows). The three artistes on stage at that moment are frozen into a group of the utmost despair, a group that might have been fashioned by the hand of a classical Greek sculptor. Such immobility is a new idea in opera, where continuous movement has been the norm.

Even the *dirigeant*, Signor Andelli, may be considered as part of this tableau, he stands as though exhausted, his arms by his side, his head bowed. The audience permits itself the luxury of a sigh. Nothing else is possible, though there was some discreet use of one or two delicately-laced — and no doubt perfumed — handkerchiefs.

A new star has arisen in the musical firmament, a star which shines brightly and explicitly. Signor Andelli has neither Wagner's woolliness nor Puccini's declamatory nonsense.

A completely new idea in music-drama . . . .'

It was not often that Hohenberg, feared and hated as he was by musicians of all sorts, set the seal of approval so firmly on any new work. Like all critics, he normally ignored or detested anything that smacked of experimentation or newness. It can only be supposed that on this occasion he had dined well.

In spite of his unwonted generosity singers, musicians and audiences have agreed in principal with his judgment. Only his fellow critics perhaps had expected some other astounding manifestation of originality than a period of silence after an unresolved diminished 7th, since the value of silence is not appreciated by the generality of the critics of music. Uncomprehending, they accepted Hohenberg's *ukase*. The craft of criticism has its loyalties no less than, say, politics.

There is a great thankfulness in Andelli's heart. He is quite aware that his music has qualities that set it apart from other music. He is also aware that he is answerable to no-one for the result of his effort. But he also knows that without Isabella Weber to sing it, it would be nothing. If she had not recognised his possibilities as a composer, and if he had not recognised that hers was a voice without equal, there could have been no *Alma Perdita*. He has since written other music; as a composer of weight he is respected; but her voice was the most important of the instruments in his repertoire. The discipline imposed upon him while writing for that voice had ultimately produced a work of genius and precision.

Out of his love for the young singer he had written his opera; in her gratitude she had sung it. In the course of one evening, an evening of fearful anticipation and trembling, both artistes were firmly established.

On his podium, Andelli lifts his eyes and his arms.

The statuary on the stage melts imperceptibly, resolves into separate and individual figures. Behind in the shadows, the chorus weaves into a semicircle of compassion.

The bows of the violinists, moving as one, stroke the strings gently, singingly. The chorus comments, pianissimo, on the situation, strophe and antistrophe. A sudden crescendo denounces the past, commends the future. The Abbot's great voice insists that Cordelia would be well advised to do the same.

'No, no!' she cries. 'It cannot be.'

The past is reality, a part of her life, bad though it may have been. The future is probability. The past is memorable, something one has fought and vanquished; the future is a dark and terrible enemy.

Crescendo, but still persuasively *cantabile*.

The violoncellos have a minatory accent. The body of the chorus repeats its commendations, to which the maid Marella adds her pleas in a still voice.

Once again supported by the strong arms of the Abbot, Cordelia looks around her in fear. She breaks away. With outstretched arms she thrusts the future from her.

'It is better that I should die, for I have sinned.'

At the word *peccavi* a shiver runs through the House — auditorium, stage, and backstage. It is the word upon which the music modulates into the introduction to the last Aria of Act I, the obstacles over which many a proud singer has stumbled, never to recover.

The hypnotic compulsion of the situation is expressed in the relentless tempo of the double basses, the agitated staccato of the 'cellos and violins as they soar upwards.

In the Royal Box the Ladies and Gentlemen-in-Waiting are neither whispering nor giggling. His Majesty sits as though turned to stone; Amelia Viktoria, even, is plumply immobile and strangely silent. Her mouth is slack, her fan languid in her lap.

Andelli seems to be possessed; the demoniacal glitter in his eyes is perceptible even from the wings where Christopher stands, and Maria, offering a silent prayer. Behind her is Hans, on almost exactly the same spot where the child Isabella had once stood. His cap is in his hand, in silent worship; there are tears in his eyes. His lips move: *I shall sing, Hans . . . .*

The arpeggio soars to the note on which Cordelia is to begin her Aria. The playing of the music takes a few seconds only, but it expresses the bitterness of a lifetime.

Lisa shrinks away, as though expecting some terrible mishap; the Abbot stands silent, unable to plead more. There is not a movement in the chorus. Cordelia moves a little downstage; the spotlight follows her precisely and smoothly, as though its mechanic were meeting a challenge also. In the brilliant mote-laden beam Andelli's baton flicks like a lizard's tongue.

Ah. Non posso far,' signori . . . . The notes glide smoothly, effortlessly, over the solid silence on the theatre.

'Ah,' sigh the Ladies-in-Waiting.

In the wings Christopher takes Maria's hand, and presses it in happiness and relief.

'She has done it, Maria. She has done it.'

'*Gott sei dank*'. But there is no reason why she should not.'

Hans, who has really no right to be there at all, twists his woolly cap round and round until it is quite shapeless. He nods happily to anyone who may look his way. 'Our Isabella,' he whispers with pride.

A little smile touches Andelli's lips, and goes. He nods approvingly. The orchestra as a body perceives the smile and the nod; it feels a new challenge; it plays with such precision that it might be one man.

There is one exception. The percussion player is excluded by the nature of his instruments. He is an individualist. Yet only the knowledge that, with all the orchestra at his command, the dirigeant's sudden glance is for him alone sustains him in the contemplation of this doubtful claim to individuality.

He waits.

He counts.

He raises his beater.

A flash — no more — from Andelli's eyes releases his tension. He flicks his triangle delicately, almost disdainfully. The diamond-edged vibrations point Cordelia's liquid notes. The percussion player sits back with folded arms; he considers the perspiring agitations of the orchestra with a kind of pitying Jovian remoteness.

As though to mimic the triangle's offering, the travelling clock in the dressing room announces the hour to the ghost-filled room. It plays through its air without faltering. It has a clockwork heart, and does not falter.

It is nine o'clock, and the snow falls silently in the square.

'Music is an exacting mistress, Isabella,' said Herr Smith, but whether it was to prepare the ground, or to excuse future tediousness, he was not certain. As for Isabella, she soon became aware that learning to sing seemed to have little to do with singing. She, who had breathed and sung as naturally as a bird, now discovered that these hitherto instinctive actions required thought and concentration. She breathed in and out; walked, stood and sat according to instructions, like a dog obeying its master's voice. These actions she repeated until she was quite bored. She could understand no reason for them; they had nothing to do with singing, she considered, because she was specifically forbidden to sing.

Daily she stood before her looking glass, breathing in and out according to Herr Smith's instructions, her hands finger point to finger point above her stomach, the better to gauge the expansion of the diaphragm. Staring at her image, she was intrigued to discover that it was possible for the human face to take on a variety of expressions. These could be classified — though not so precisely by Isabella — as the ugly, the normal, and the beautiful. She was more adept at the ugly; indeed, in her mind the ugly was normal; the red hair, the green eyes, the far too large nose . . . . Whatever Beauty was, she did not possess it. She frowned, she pouted, she smiled. All of no avail.

Herr Smith watched her occasionally at these exercises with some amusement and affection.

'Beware of vanity, my child,' he admonished her.

'I do not know what vanity is.' She faced him defensively, sensing a criticism of her actions she did not understand.

'Vanity is a great evil; it creeps into the mind and the heart, and possesses them entirely.'

'Why am I not beautiful?' she turned upon him defiantly.

He studied her in all seriousness for a while.

'Alas, Isabella, what is beauty? If you are thinking of the shape of the face merely, that is scarcely beauty. Rather it is an accident of nature. We can do nothing about it. In the end beauty could be, I suppose, the sum of the good qualities of the human being. Some call it "character". That is the good side of it. Personal, physical beauty, although very pleasant to behold, is a snare and a delusion. Like the seven sins it destroys those who have it, in the end.'

Isabella pouted.

'I would like to be beautiful, all the same,' she said.

'Of course you would. You are very pretty. Whether you will be beautiful or not remains to be seen.'

With that she had to be satisfied. Herr Smith made little of it, except that he was compelled to recognize the fact that Isabella would one day be an exceptionally handsome woman, if not a beauty in the accepted sense. She was growing taller, and the tragic hollows in her flesh were filling. In deference to her budding feminine psyche he allowed her the little vanities she paraded, the brief periods of sulking, the occasional flashes of temper. He gave her the best food and wine that he could afford. He became friendly with Madame the *Modiste*, whose establishment was in the Square. She ran up sensible dresses for a growing girl, and 'confections' for what she described as 'A young Miss'.

He was proud of Isabella. She offered him her childish love and trust, and the bonds between the two grew stronger. There were occasions when he was made uneasy by a momentary demonstration of a quality which could have temper, stubbornness, or perhaps singleness of purpose. It was difficult to find a reason for these rare exhibitions, and happily they were always, although explosive, brief. Herr Smith was unhappily convinced that Isabella was beginning to think the world revolved around her.

He coped with this situation in his own way. She was still a child, he persuaded himself, and her short life had been hard and ugly. She had much to forget, and more to learn, and she could be forgiven if her mind was confused and uncertain. Perhaps a tight rein might be necessary for a time, he considered, thereby demonstrating the inadequacy of the masculine mind.

In his less than adequate way he continued and increased the exercises he had begun in the firm belief that if she were driven to exhaustion by physical extremes she would be less inclined to give way to instincts and impulses. More and more he began to think of her as an instrument to be perfected to the limit of its capacity, and until that perfection was attained there could be no place for exhibitions of temper or of temperament, no time for wishing to be beautiful, no giving way to the weaknesses of the flesh.

Because he loved the child, and because his own character was less than forceful, his methods were only partially successful; and when sometimes temperament did gain the upper hand for a time, he thought it was a pity that her soul was not quite the right shape.

*

The days passed into weeks and months. Isabella's time was as fully occupied as it could possibly be with agitations she still considered as having little to do with singing. She was sometimes extremely bored, but there were other diversions; they walked about the town and the countryside; they talked about all manner of things as though they were two children. He also began to teach her something of the theory of music and voice production (which he prepared secretly beforehand from a heavy German tome he found in a dusty bookshop). For some time Isabella did not actively rebel.

Herr Smith perceived that what he liked to think of as the first part of his training was beginning to bear fruit. The road was to be long and arduous, but he was secretly satisfied as he noted how she composed and comported her body in effortless harmony; how she arranged her arms and hands in positions in which simplicity disguised artifice. Already she manifested an awareness of the importance of self-control and discipline which was to be the public image of Isabella Weber.

In an age in which temper and temperament were the norm, she was always to be calm and self-controlled in her art, and the opera-loving public, surfeited with the lions, the tigresses and other fauna of the musical jungle were to take her to their hearts for this quietness.

Sometime towards the end of this confused apprenticeship Isabella was allowed to sing again. That is to say, on every seventh day she was taken through a series of scale-like passages designed to explore her capability, then to extend her range, then to improve the tone. Some of these passages were culled from the tattered operatic score books reposing on Herr Smith's shelves. For accompaniment he occasionally played his violin.

Isabella loved them. She was as enchanted as if she had indeed been performing in grand opera.

'At last!' she exulted. 'I am about to sing. One day I shall be a great singer, like Jenny Lind. People will come to hear me; they will clap me, and bring me flowers. They will give me money. We shall be rich. You shall ride in your own carriage, and have a lackey to carry your violin.'

He smiled wryly.

'I am not sure, Isabella, if that is the whole purpose of learning to sing. Dreams are all very well, but they lead to false conclusions; besides, they use energy that might with profit be used in other ways. Practice is of more value than dreams, it has more tangible results. When you can sing a scale as if it were a prayer, *then*, you may say ''I am beginning to sing''.'

She pouted, but returned to her exercises.

Nevertheless she found some encouragement in what he said, and practised with a will. She sang scales and runs in a variety of disguises — all poured from her throat as easily as a bird sings, legato, staccato, piano and forte.

The people in the rooms above knocked no more. Passing up or down the stairs they said: 'It is to be hoped that the little nightingale does not break her heart because of such happiness.'

From time to time they stopped to listen, their fingers wagging solemnly as if they were all dirigeants, and Isabella indeed a *Prima Donna*.

There was, however, no sign of a breaking heart as yet. There was indeed no reason for her heart to break, unless it were that she could not progress fast enough.

Thinking to assist her further, Herr Smith spent a large sum of money on one of Herr Maelzel's new-fangled *Metronomes*, proudly carried it home, and proudly demonstrated it to Isabella. Thinking it was a new toy, she was at once taken with it, and enjoyed weaving a diversity of musical patterns on and about its solemn tick. She could not imagine any other use for it. Papa Smith used it every day during her exercises. It ticked humourlessly as she sang her scales, chopping off little bits of time, tossing them to her with a deliberation and a smug monotony that drove her to the verge of fury.

Eventually she broke down before its relentlessness. Her face was white, her eyes blazed.

'I hate it,' she stormed. 'I hate it. I hate it. Such a stupid box, with a noise like a clock. I hate it. It has not even a face.'

Papa Smith remonstrated unhappily.

'You should be proud to possess such an instrument, Isabella. Not everyone has a machine like this to help them count.'

'I can count. *Eins, zwei, drei* . . . . There, you see.'

'That is indeed well done,' Papa Smith complimented her with some irony. 'But not even angels could sing and count at the same moment. When one sings, each note of the same kind must be of the same length.'

'So you have told me. And it is not necessary for me to count. I know when a thing is right. I feel it within me.'

She sang a scale several times at different speeds, and Papa Smith was compelled to admit that as far as he could tell, all her notes were, according to the speed at which she sang them, of precisely the same duration.

Isabella was triumphant.

From that time on the Metronome stood silent upon its pedestal, waiting until an obvious fault in timing should appear. None did, and Papa Smith recognised with some reluctance that not only did Isabella have perfect pitch, but a perfect sense of time and rhythm also. He called himself a simpleton for not recognising this accident of nature earlier, but then excused himself on the grounds that perfection, whether of nature or art, is rarely recognisable at first sight. In his search for perfection, he had ignored the first evidence of it.

More warily, humbly even, he continued with his idea of teaching Isabella to sing. He now knew beyond doubt that his knowledge of the technique of voice production was seriously inadequate, and he was fearful lest so beautiful an instrument should be damaged by his maladroit handling. It was borne upon him gradually that his only part in the future was to encourage, to care for. The rest would follow.

Nevertheless he was very pleased with himself for having discovered the little Isabella in the first place.

*

The boundaries of music spread wider. The new and magnificent Opera House was opened in Vienna; Madame Jenny Lind, still singing beautifully, reached the age of forty-nine; Isabella Weber reached the age of fourteen. These events were only connected in that Jenny Lind remained Isabella's idol, and Herr Hans Richter, directing the Vienna Opera, was to seek out the services of Isabella Weber for his theatre in a few short years. Meanwhile, the Autumn of another year came round, with the routine opening of the Opera, the pressure of its rehearsals and the glory of its performances. Isabella practised by day; by night she was always there with Hans in the little room, its enveloping warmth, the coffee-pot bubbling in perpetual motion, the fading portraits on the walls. Always there were exotic singers and music makers passing the door, each one greeting Hans according to their nature. With a few exceptions Hans greeted them in return, according to his nature. Always there was the rumbling and squeaking of scenery being moved, and the sound of hammers making it fast.

When Isabella marvelled at all these things, and dared to comment on the god-like and beautiful creatures who were briefly within her sight, Hans so far forgot himself as to wax ironically on the subject of musicians in general and opera players and singers in particular.

No different from the rest of us,' he said. 'They only think they are different; though by what right remains a mystery that will never be explained.

Those musicians, for example, those oboes and bassoons, those violas, ten-a-penny fiddlers . . . they all think the orchestra was made for them; yet if one of them breaks a leg, what happens? A substitute walks in. There is not one of them but has not had his dreams of becoming a *virtuoso*. If they really could do anything with their instruments, they would go and do it. Why do they not? Because there is safety in numbers. Here they have the company of their kind; mutual encouragement; the discipline imposed by ensemble work. Above all, they have some kind of financial security. They join the orchestra; they stay in the orchestra if they are lucky, until they can play no more. Night after night they scrape away and bang and blow at their Beethoven and their Rossini and their Wagner until they are heartily sick of it. In the meantime they tell each other how they will take the world by storm when the opportunity presents itself.'

'How many of us would do differently?'

'But they are harmless enough. Only now and again one breaks out, probably maddened by boredom. The singers, now . . . . The chorus are the worst. Failed soloists, most of them; or they do it for a hobby in winter. As for the principals, well. There's something about an opera singer. Where they have sung; where they are going to sing next; what they have sung, and how well they sang it. How wonderful they are, how many lovers they have had. They are so conceited you would think the Sun would hide its head in shame. Male or female, it does not matter. The men are as bad as the women.'

He broke off as a head appeared at his window.

'Good evening, Madame Hausmann. No letters today, Madame Hausmann.'

'I am disconsolate, Hans. What a pretty child you have there.'

'Yes, Madame, thank you, Madame.'

'. . . German manners are not good enough for them now,' Hans continued his diatribe. 'They forswear their birthright to appear cosmopolitan in French, or Italian, or some other mincing mode of speech and behaviour; *chérie* this and *chérie* that; pawing and petting and kissing. Faugh . . . .'

'How beautiful she is,' said Isabella.

'Eh? Oh, that. The good Madame Hausmann. The present *Prima Donna* of Strelsau (for one season only, you understand; you should hear her high notes). I remember when she was younger. You could see her face then; now it is hidden beneath the layers of paint and powder.'

'Does she sing well?'

'She still sings well enough, except for the high notes, as I said. But already they have had enough of her.'

'Will she sing tonight?'

'She sings every night. There is no stopping her. She has lost the better part of her voice straining at Wagner, and other excesses. It doesn't do, in my opinion. Special training is necessary.'

Thus Hans pontificated, while he made coffee, kept a sharp eye on the stage door, and exchanged greetings with those who entered, stamping their cold feet.

'I am told you are beginning to sing,' he remarked suddenly, turning that same sharp eye on Isabella.

Isabella confessed that she thought she was.

'Ah,' said Hans, as if that were the end of everything. 'You always said you would sing, as I recall.'

'You think I should not sing?' Isabella faced him squarely. 'But I want to sing, more than anything else I want to sing.'

Hans sighed, and shrugged his shoulders.

'If you want to sing, you will sing. That is certain, for nothing else will matter. Very well, then, sing if you must. But understand, singing is no easy thing, these days. Oh, singing itself might be easy enough, once you have trained. It is breaking into the ring that's difficult, however good you are; getting people to listen to you, and getting them to like you when they have listened. And what is the end of it all? The theatre. The wickedest way of life man ever devised. The theatre is a man's world, and only a man can find a place for you in it. At a price . . . .'

He stared at her for a moment.

When he spoke again it was with great sadness.

'When I look at you, I think it does not matter whether you sing well or ill. Either way I see trouble.'

'Papa says that I sing very well,' Isabella told him, with some spirit.

'Then,' Hans assured her, 'it must be so.'

What was to be, was to be, he thought. It was no business of his. However much he grumbled about it, the world she proposed to enter was no world of his; his pontifications to Isabella about this and that of the world of opera were a jumble of what he saw and what he heard, and bore little or no relation to the truth. As she stood from time to time in her corner of the stage, she saw and heard a magical world within a world; he saw a dusty stage, make-believe scenery, ugly, ill-fitting costumes; people vocalising, for ever striking attitudes. Isabella drank in the wonder of it all. She mouthed the words as they were sung, or hummed the music; she watched the puppets as they moved, and registered every gesture in her memory.

How well she learnt by merely listening and watching was demonstrated when she imitated the good Frau Hausmann at work, emphasising the gestures and even the vocal mannerisms with a fidelity that was little short of cruel.

Herr Smith now decided that the situation was beyond his control.

*

There was in Strelsau only one other possible opinion — that of Herr Josef Weingartner.

Herr Weingartner and his sister Matthild' — who kept his house for him, played hostess to his gregariousness, and accompanied his students whenever he allowed them to sing anything worth singing — lived in one of the pleasantest corners of the city. The house was Josef's by the grace and favour of the Grand Duke Michael (who found the greatest satisfaction in being generous to musicians of every degree and of either sex); it was low, steep-gabled, and white, and it had green and red shutters. It occupied a corner on the southern side of the great Stanislausplatz. The salon was furnished with three large windows; from the window on the left a view was to be had of the facade of the Opera House; from that on the right a perspective of the Royal Palace (a show-piece only, since the Family preferred the isolation of the Castle outside the city). Through the middle window was to be observed the imposing pile of the Cathedral of St. Michael. One had therefore, almost without the pain of disturbing oneself, a complete integration of the sacred and the profane. It was unfortunate that some inattention to the siting of the equestrian statue of the Grand Duke Stanislaus (great-grandfather of the present monarch) had caused the rear aspect of that prince's noble mount to fill, rather vulgarly and permanently, the middle panel of this triptych of windows also.

It was not necessary, as Herr Weingartner frequently pointed out, with lack of delicacy and gusts of laughter, to look at *that*.

Beyond the statue, on the farther side of the square, rose the grey stone walls of the prison. This monument had been the pride of former Elphbergs, who held strongly the belief the punishment should fit the crime. Of late, some attempt had been made to soften its threatening outline by the cultivation of a row of plane trees. It was far enough away to be picturesque.

Continuing the frontage of the house, and thence to the side and back, a high wall enclosed a courtyard in which grew a variety of herbs and seasonings, and along the wall which received the heat of the sun a majestic vine, Herr

Weingartner's pride and joy. It was his custom to sit in this courtyard, in the cool of a summer evening, in the company of his sister; he with his pipe — a particularly vile 'Churchwarden' — and a tankard of beer; she with the needlework which was her preoccupation when she was not at the forte-piano.

Herr Josef Sigismundus Weingartner was known to all Strelsau as the Herr Kapellmeister Weingartner, but he was presently Assistant Court Secretary and Singing Master. His passion in life was singing and the teaching of singing. He frequently made it quite clear that there were few better teachers of the art than Herr Weingartner, for modesty was not one of his faults. When he was not teaching, he was human; when he was teaching he was an ogre, a torturer, heeding neither tears nor entreaties; accepting neither failure nor excuses for failure. He began by frightening his pupils; he continued by inspiring them; he ended being loved by them. In spite of his ferocious discipline and his caustic tongue, he enjoyed the *entrée* to the best Houses of Strelsau. The ladies protested that they did not wish to hear his version of the latest gossip, nor to hear his account of the exchange of courtesies between sensitive songsters; nor to learn who was and who was not in favour at Court at this moment. The gentlemen were jealous to a man of his success with their wives and daughters.

Herr Weingartner told his tale with such wonderful good humour and just the faintest touch of malice, that no one could be offended. Since musicians were expected to be quite other than moral, he invented a kind of immorality for the ladies that hurt none of them. They were enchanted at his innuendoes.

'Dear Herr Kapellmeister,' they would exclaim, fluttering. '*Lieber Herr Weingartner*,' they would exclaim. 'So droll . . . . *Also komisch* . . . .'

They would smile at each other knowingly.

They would have been more than surprised to learn that he had a very low opinion of them and their pretensions, and that quite simply he was using them, although he too enjoyed the *recontres*. They were rich and influential, and by manipulating them judiciously he was able to direct the progress of many of his less well-endowed but more talented pupils. Having taught their daughters to use what little voice they had, he was thereafter enabled to help those who were more deserving.

To Herr Weingartner Papa Smith brought Isabella.

If he had doubts about approaching so important a person, he brushed them aside. He was conscious of his own lowly state in the musical society of the city, and expected nothing more than advice to take the child to some back-street teacher of singing.

Papa Smith was surprised to find that Herr Weingartner was an easy person to approach. He received them without moving from his chair by the window, gazing benignly from one to the other, puffing at his long pipe with every sign of contentment.

'So this is your little song bird, my friend,' he rumbled between puffs, in a voice like distant thunder.

He studied Isabella as she stood demurely before him, scrubbed and immaculate. He noted with approval her black dress and white belt, and her unruly red hair now tightly disciplined by a new green silk riband. He smiled happily as she curtseyed to him and to his unsmiling sister, who was occupied as usual in the fabrication of a complicated design embodying the most unnatural flowers that could be imagined.

Isabella returned his gaze openly. She did not know whether she had ever seen a larger gentleman. He was indeed, if she had known the word, abundant. He completely filled his canary-yellow waistcoat, stained with tobacco ash; he filled his purple coat and knee britches; he filled his yellow hose and his great silver-buckled shoes. In his features he resembled a larger Herr Schubert (of whom she had seen pictures), but unlike Herr Schubert's, his face seemed to be at once kind and severe. Isabella thought that perhaps he made himself appear stern so that he should not smile too much. His voice fascinated her; it was a voice full of laughter and sympathy. It was the voice of a man who lived a full life, a life of his own choosing; it was the voice of a man who was master of himself, and therefore master of others, whether he would be or not.

Isabella could not, of course, comprehend this immediately. She was afraid of him, and at the same time she loved him.

Neither emotion showed on her face. She stood before him unmoving, her hands clasped lightly in front of her. As he studied her, Herr Weingartner was aware of a faint stirring of unease. There was a familiarity about the face he could not place. With certainty he knew that he had seen its like before, not once, but many times, and he was angry with himself because he could not recall in what context.

He had expected — after sorting out Herr Smith's confused explanation of why he wished to consult him — to be confronted with a neglected, perhaps orphaned child with some quality of voice capable of attracting a simple old man. It was all wrong. Herr Smith was neither simple nor old, and the child was far removed from being a neglected orphan. Herr Weingartner had expected a child of whom he could say: *Yes, she sings very prettily. Very prettily indeed. Bring her back in three years' time, and then we shall see*, and then forget the matter. Confronted with this child-woman, who was tall and slim, and gave promise of beauty, and who stood before him proudly, if a little apprehensively, Josef knew neither what to think nor what to say. He recognised only at this first encounter that those green eyes, that red hair, that perfectly moulded (but somewhat large) nose, that firm chin indicated a decisiveness of character that contrasted oddly with the youthful figure.

He laid his pipe down, sighed gustily, and blew his nose into a handkerchief as big as a small tablecloth.

'However she sings,' he thought, 'here is a pocketful of trouble for someone in the not-too-distant-future.'

He indicated a raised part at the end of the room, whereon stood a forte-piano.

'Take your place there, Isabella Weber,' he instructed her. 'Turn towards me, but look at the picture above my head.'

The picture was an oil painting entitled *The Music Lesson*, a gift from one of his more grateful pupils. As Isabella took up the place he had indicated, he noted the grace of her movements, the controlled poise of her body as she stood facing him, and the proud lift of her chin. Again he was perturbed by the shadow of a memory, against which he struggled vainly. Who was this child? What was she? An orphan, adopted and cared for, Herr Smith had told him.

If her voice were to be as striking as her physical presence here was a treasure indeed. But the question still persisted: *Who was this child*?

From her place on the platform Isabella looked over the tight glossy head of Fräulein Weingartner bent in concentration over her needlework; over the abundant greying locks of Herr Weingartner, at the picture of *The Music Lesson*. The afternoon was advancing; the sunlight entered the room through the garden window, and fell upon the picture, upon the brocade upon the table beneath it, upon the bowl of fruit. It transformed each object in turn to gold, as if by the touch of Midas himself.

The voice of Herr Josef Weingartner came to Isabella through the gold, a gentle voice, quite unlike that which had struck fear into so many recalcitrant pupils.

'You may begin in your own time, Isabella.'

Isabella, a sudden wave of nervous agitation upon her, glanced in desperation at Papa Smith. His eyes were tender; he smiled at her and nodded.

'My child,' he had told her, 'you may have a headache; you may feel ill; you may think the whole world is a terrible place. It sometimes happens. Nothing matters but that you should sing. When you have sung your first note, all that will vanish, for you will then be in your own world.'

To Herr Weingartner she said: 'I will sing the aria *Ah, Fors e lui.*'

She began perfectly, *legato* and *piano*. At the first note her fear slipped away. She was no longer a fearful would-be pupil, she was a singer, singing for pure joy. The difficulties of the piece served as a challenge to bring out the purity of tone and the suppleness of her voice. The propriety or otherwise of so young a child assuming, if only temporarily, the mantle of a Parisian demi-monde, occurred to none of them. The words at this time had little meaning, the voice was all. To himself Herr Weingartner considered that Papa Smith had prepared the ground — he smiled slightly in Papa Smith's direction — with sensibility. Herr Smith, he thought, was an excellent musician, without doubt; his training of the child so far showed sensitivity and feeling. A word in the right direction might be of great help to him.

Isabella continued to sing. Fräulein Weingartner paused between two stitches, and regarded her brother over her spectacles. Her face had lost nothing of its accustomed severity, but the pause was meaningful. Josef placed his fingertips together judicially, and nodded gently. His sister was notoriously difficult to please. If she approved of this child, his unspoken judgment was confirmed, and his way clear.

The change to the major key with all the attack of an experienced opera singer took him by surprise. He perceived that the child herself had no notion that she was singing quite effortlessly a piece that many had failed to sing satisfactorily after many years of training. She was, quite simply, singing as she had been requested to do, and Herr Weingartner was privileged to be her first auditor, to experience what many others were to experience, to their great surprise, when they later heard her.

That evening he wrote in his day-book: *My skin horripilated; my forehead perspired. Other singers have caused this to happen, but not to the same extent, nor in my own salon. They had the emotional atmosphere of the opera house to help them to cover their mistakes. Here is a singer who makes no mistakes. Here is a singer, young, without experience or knowledge, as I understand, whose voice is naturally endowed with a roundness, a richness, a fine susceptibility to control which will, in my opinion, make her the leading soprano of her time, if not of all time . . . .*

Madame Mathilde Marchesi, reading in Vienna this passage transferred to Isabella's letter of introduction from Herr Weingartner, smiled a little wryly. Dear Josef, so many of his swans had turned out to be merely ugly ducklings after all. But that was before she herself had heard Isabella sing; after hearing her, she knew that he, who sometimes allowed himself to exaggerate where his pupils were concerned, had not exaggerated on this occasion.

She noted in her turn the perfect pitch, the perfect rhythm, and the fact that every note was attacked squarely and sung fully. Isabella was never guilty of *feeling* for a note, or of hovering round the edges of it. After making allowances for dear Josef's enthusiasm, Madame Marchesi came to the same conclusion as he had, but with some reservations. Nine-day wonders were all too frequent. Meanwhile, there was little more to be done except to wait for the voice — and the personality — to settle in its own good time. A period of rest was necessary, to be followed by a gradual musical re-education.

Yet, she wrote officially to Herr Weingartner after she had accepted Isabella as a pupil, *lest the child should feel that she is not making progress, I will continue with all branches of musical study and vocal art, gently at first, and after that, who knows? The world may well be at her feet . . . .*

*

The aria came to an end.

The singing birds in the cage hanging outside the open window began their interrupted trilling as if in jealous emulation of that golden voice. By the forte-piano Isabella waited, calm and composed, breathing normally. She had sung, as she had been asked to do. What was to happen next?

Fräulein Weingartner was no longer preoccupied with her needlework; it lay idle in her lap. Her fingers were clasped tightly together, and her eyes were fixed on a point which lay far beyond the confines of this pleasant room. On each sallow cheek a tear hung trembling. Herr Weingartner blew his nose with more than necessary force. He wiped his brow and sighed. His sigh was itself a miniature gale.

'Incredible,' he said, after the demonstration. 'Incredible in one so young.'

He heaved himself out of his chair, and crossed to the forte-piano. He arranged a sheaf of dog-eared manuscript on the rack, and flexed his fingers.

'Come, Isabella, sing to me from these pages. Can you read music?' She peered over his shoulder at the spidery marks below a florid Italianate signature which almost filled the top half of the page — *Rossini*. 'I think so. It will be difficult. The writing is not good, I think.'

Josef roared with sudden laughter, slapped his thigh.

'Do you hear that, Matthild'? Rossini's writing is "not good". How my old master would have loved to hear his handwriting described by a chit of a girl as ''not good''. These passages, Isabella, were written in 1850. The Master was then ill, in great pain. But he wished me to have the best, the most exacting of exercises. I did not fulfil his expectations as a *tenore*, but because he loved me he gave me the manuscript you see here. It may be that he foresaw, in his pain, a greater use than I was able to make of it. So, Isabella, it may also be that that time has come. Let us sing the Master's Workshop Music.'

He smiled at her, and she was relieved that he smiled rather than laughed again. There was something overwhelming about his laughter, and she felt that she had offended in some way. He, however, showed no sign of being put out. He stabbed at the page with a plump forefinger, and struck a chord, initiating a series of modulated *arpeggios*. Isabella sang, not without difficulty at first, but when he played the final chord, she was exactly on the note.

Turning the pages, he indicated a passage of chromatics and plunged immediately into it. Isabella followed without hesitation. Another page — trills, turns, thirds, broken chords, skipped octaves. In a short time Isabella had mastered Signor Rossini's not-so-good-writing. But actually she had ceased to read. Her own musical sense had taken up the score, a point which did not escape Herr Josef.

'Incredible,' he said to himself again. But he appeared to be satisfied.

He rose from the instrument.

'Enough,' he said. 'These are but parlour tricks.'

He returned to his chair by the window, and motioned both Isabella and Papa Smith to accommodate themselves in other chairs. Fräulein Weingartner took up her needlework and continued to fabricate her monstrous flowers as though nothing out of the ordinary had happened. But Isabella was uneasy. She could not understand the silence, broken only by the singing birds, that had fallen upon these three people. She had sung quite happily, as she always sang. If they wished to be serious about it, that was their affair.

Since they had not finished being serious, she waited patiently.

Herr Weingartner filled his pipe carefully; tamped down the tobacco, and applied a flame to it. He puffed for some time, until he was enveloped in a large cloud of smoke. He appeared to have come to a decision.

'Herr Smith, why did you bring this child to me?'

'You are the best teacher of singing in Strelsau, Maestro, possibly in all Ruritania. It is well known.'

'Thank you, Herr Smith. Compliments apart . . .?

'I did not intend a compliment, Maestro. Only you could tell me whether the voice is as good as *I* think it is.'

'Good! What is good? The child sings easily and naturally. She has a present range of two and a half octaves. The tone is firm and mature . . . .'

He stopped suddenly.

'Mature. Did I say mature?'

Fräulein Weingartner spoke for the first time.

'You did, Josef.' And the tone of her voice indicated the measure of her distaste for what she considered to be another of Josef's exaggerations.

'Then I must have meant what I said,' replied Josef unperturbed, 'unlikely though it may seem. What more do you want, Herr Smith?'

Herr Smith sought for words to express what he needed, and could find none.

'I am not myself skilled in the teaching of singing, Maestro. Perhaps I have done wrong. Perhaps I have attempted too much, strained the voice . . . .'

He trailed off miserably.

'Your training has been satisfactory, my friend, considering your admitted limitations. But this child . . . .'

Herr Weingartner paused. He wished neither to raise false hopes, nor to lay himself open again to a charge from the taciturn Matthild' of exaggeration. Nor dare he allow himself to recognise that here at last was the voice he had waited for throughout his teaching career, a voice, fortunately allied to a promise of physical beauty, that he could eventually present to a world which assuredly never before heard such perfect sounds issue from the human throat. The voice was now his, should he wish it, to work such a miracle with. He could not allow himself to recognise that there was little he, a mere man, could do except fill the role of caretaker. As far as the act of singing was concerned, Isabella was a finished product.

Following the changes of expression on his face, Papa Smith plucked up sufficient courage to ask:

'What can I do, Maestro? What are the possibilities?'

Reluctantly Herr Weingartner allowed Papa Smith to re-enter into his dream.

'Do, my friend?' he asked. 'Do? We can do nothing. Possibilities . . .?'

Papa Smith's face fell.

'. . . the possibilities are endless. You — and I, I may say — have a valuable property on our hands. We are not looking for child prodigies, there have been too many of them already. We are looking for good, lasting durable material. It is a matter of marking time while we maintain the present standard. The child is still no great age. She is not ready to astonish the world nor,' he made a wry grimace, 'is the world ready to be astonished by her.'

'But, Maestro . . . .'

'Have no fear, my friend. Nothing will be lost, and much will be gained. Nothing new must be attempted in the way of singing, but there are still many things to learn — some theory of music, for instance. Some history of music, the lives of musicians (with one or two reservations). The child must also learn to play the forte-piano, a valuable accomplishment in itself for a singer.'

'All these things are beyond me,' protested Herr Smith. 'I have little knowledge of the lives of musicians, except that I play their music on my violin. As for the forte-piano . . . .'

Herr Josef Weingartner thrust up his head in simulated anger.

'Dolt,' he almost shouted. 'Do you understand nothing? From this moment on she is my responsibility in the matter of singing, and the rest of it. My dear sister there will attend to the forte-piano lessons. Are you now satisfied?'

'Do not think, Maestro,' stammered Herr Smith, 'that we are lacking in gratitude when I say that such a course was far from our intention at this time. We desired an opinion only.'

'You have an opinion, friend, and a decision. In fairness to the child herself the matter can scarcely rest in the giving of an opinion. She must have her chance. I can give her the chance. In twelve months' time, when matters have been arranged, she shall go with my recommendation to Vienna to Madame Marchesi. I have already decided.'

Herr Smith gaped. It seemed that the ground was slipping from his feet.

'Maestro,' he said at last, 'that will cost money, a great deal of money. I have only my salary from the opera, and a few pupils.'

Josef held up his plump hand. He might have been a bishop bestowing a benediction.

'My sister and I will undertake her further musical education gratuitously. As for Vienna and Madame Marchesi, there are those who will be glad to underwrite the proposal when I have put it to them.'

As an afterthought, he added:

'I will also arrange that she has access to a forte-piano.'

The forte-piano in fact came to Isabella.

It arrived in triumph, riding high upon a cart pulled by two mettlesome horses. An interested crowd of neighbours elicited the information that it was destined for Herr Smith.

'That will be for our little singing bird,' the neighbours were voluble. A score of voices offered advice as to the best way to get it up to the third floor. The three strong men who had accompanied the instrument ignored the confusing and sometimes raucous advice, and went self-consciously about their task of hauling its component parts up the stone staircase to Herr Smith's apartment. Finally it stood, sleek and shiny, in the middle of a room scarcely big enough to contain it.

When the tuner — one of the three — played his first chord upon it after its assembly, a roar went up from the street below.

From the first moment, Isabella worshipped it. She could not caress enough its inlay of mother-of-pearl and ebony, its sleek shiny body painted with representations of goddesses making music, its diatonic keys as white as snow, and its pentatonic keys as black as night. There was no indication of its origin, and when Isabella, in all innocence, asked Herr Weingartner whom she had to thank for such a noble gift, he replied shortly: 'The Friends of Music, let us say.'

And that was all.

Being little more than a child, Isabella took the situation for granted. Daily she sat at the keyboard; twice every week she almost ran, so eager was she to conquer this new world of music, to the house in the square where Fräulein Weingartner undertook with some graciousness her education in forte-piano and theory of music, and struggled to lay a foundation of the French and Italian languages. Progress was rapid; Isabella absorbed all that she was taught, and taught herself more. She practised until her fingers were quite sore, until even Herr Smith was compelled to put his hands over his ears at each repetition, and cry out for her to stop and rest a little for his sake as well as for her own. She continued to drive herself almost to the point of exhaustion, and there were the inevitable times, as in her singing, when some piece of technicality defeated her temporarily, or she felt that she did not progress quickly enough. There would be rages, tears, sobs.

'Why cannot I play it?' she would storm. 'What is wrong with my fingers? Look at it. See how easy it is, and I cannot play it.'

Nevertheless, at the end of some minutes, sobs still shaking her, she would drive herself as furiously as before. Papa Smith was unhappy about it, fearing for her health. Isabella ignored him, just as Herr Weingartner appeared now to ignore her, for all his promises. If she could not sing, she would play. After twelve months she could play the forte-piano as well as most recognised concert artistes, and certainly with more ability than any opera singer or, indeed, than her *repetiteur*.

This achievement was almost wholly due to her own efforts. Fräulein Weingartner was not so musically inspired as her brother, and her playing, as well as her teaching, tended to restrict itself to demonstrations of the prettier drawing-room pieces which she herself executed daintily enough but without that understanding which made them interesting. Only rarely did she venture into the more serious works for the forte-piano, when her brother Josef demanded her services as accompanist for his pupils. At the beginning therefore, and naturally enough, there were finger exercises, the usual scales and arpeggios, the boring practice themes from sonatas, rondos and marches, so essential for the tyro pianist. Later came extracts from *Intermezzi*, Italian *opera buffa, Singspiel* from Germany, such as were loved by undiscriminating townspeople, pieces by Hiller, Muller, Weigl and Gelinek. These were crooned and whistled by both high and low in Strelsau, but Isabella soon tired of their lack of depth and sincerity. Coming by some copies of the sonatas of the great German Beethoven, and some sheets of arias of Mozart, Isabella was enthralled. This, she considered, was *music*. In spite of his *diktat* that nothing new should be attempted in the way of singing for at least twelve months, Josef Weingartner relented from time to time, and allowed her to try the Mozart. He persuaded himself that he was keeping the voice in working order, to disguise his lifelong ambition to present to the world the perfect Mozartian voice. He was perhaps a little eccentric when he insisted that anyone who could sing Mozart perfectly could sing anything. Isabella sang Mozart to their mutual satisfaction, and the reluctant singing lessons became more frequent, although not always with sympathetic *rapport* between music and pupil. Disputes arose. Concentrating on the singer, Herr Josef forgot the person. In her turn Isabella, growing up, no longer avid for advice, rejecting guidance, began to interpret. The results were frequently disastrous.

Herr Josef had a great dislike of excessive histrionics, and in particular nothing irritated him more than the pointless gesturing with hands and arms so indulged in by second-rate singers. On one occasion when Isabella waved her hands about more than he thought necessary he struck a sudden discord and stopped playing. Isabella stared at him.

'What have I done wrong, Maestro?' she frowned.

'When singing Mozart,' he told her with some asperity, 'it is not necessary to behave like a windmill. Interpret with your voice, my child, if you must interpret, not with your hands.'

'Especially when you are singing solo,' he added.

A feeling that Isabella was beginning to recognise and fear welled up within her. Nevertheless she could not control it; she stiffened, she considered herself offended. She was no longer a child; she was a singer. The expression on her face interested Herr Weingartner. Again he was troubled by it, and her prideful posture, resemblance to someone he could still not recall.

'Must I then stand always like a dummy?' she demanded. 'How could I act a part if I did not move?'

'Do not act, child. Do not try to "act". There is indeed no such thing as "acting"; it is merely the additional colour on the canvas of experience. Exaggerated gesture, on stage or off, is a sign of emotional — perhaps mental — insecurity. Stand like a dummy. Why not? Sing only. That is all I ask. The work is greater than you. Inevitably there is more in the music that a mere mortal can express. Do not try to manipulate it; let it manipulate you. In the end such "acting" and gesture as is necessary will manifest itself through the understanding of the music.'

This reprimand, delivered in the gentlest voice of which Herr Josef was capable, was a great shock to Isabella. She had thought, indeed she had been led to think, that the voice was all important, that it was the first of all instruments.

'Indeed, there is no doubt of that. But,' Josef reproved her, 'we are not now speaking of the voice.'

'Of what then are we speaking?' Isabella enquired coldly.

'We are speaking perhaps of self-importance, self-understanding,' he replied.

Isabella stamped her foot.

'Do you mean to imply that I consider myself important?'

'I mean to imply nothing save that perhaps humility is a lesson so difficult that many fail to learn it. Consequently they never sing.'

Isabella's eyes flashed ominously.

'It has escaped your notice then, Herr Kapellmeister Weingartner, that I sing well enough already.'

Josef sighed gently, almost inaudibly. He had had many such scenes with would-be singers, even with those who had, as they thought, attained their objective. The *psyche* varies little from person to person. But here there was a difference. The impatience, the flashes of temper, even the little *bravura* of stamping the foot (which, if truth were told, Isabella had picked up from her observations of certain 'actresses' as she stood in her corner at the Opera

House) were nothing. Deep in Isabella there was a pride, a haughtiness, a defiance which set her apart from others.

She awaited an outburst of anger from him, but none came. He turned his back to her, and continued to stare through the window. He contemplated the trees, greenly transparent in the summer sunshine, the violet shadows on the white rough-cast wall, the cage of singing birds. He murmured into the quiet of the morning, as though he were speaking to himself:

'*Be still, and know that I am God . . . .*'

Isabella trembled suddenly. Whether from fear of the awful words, or from anger because she was ignored, she could not have said.

Still in the small voice Herr Josef continued to muse:

'The ideal singing lesson should consist of ten minutes on the knees, with head bowed, and twenty minutes developing to the best of our ability the voice that God in His Goodness has seen fit to bestow upon us.'

A gust of anger shook Isabella. With studied calmness she drew on her gloves and took up her music. Only the slightly impatient flirtation of her parasol betrayed an inward struggle.

At the door she paused, and curtseyed, a chit with the deportment of an archduchess.

'I shall sing no more today,' she announced to Herr Weingartner's broad back.

Without moving, he replied: 'It is perhaps as well, *gnädige Fräulein.* Go with God.'

As she left the room the birds in the cage began to sing as much as to say: 'Now we can come into our own again.'

'I perceive,' said Herr Weingartner to the singing birds, 'that I have a firebird on my hands as well as a nightingale.'

It was not to be the last of such fiery exchanges, as he was to learn to his cost.

*

As on all other occasions when she was possessed by what might be called her *alter ego*, Isabella sought solitude and a temporary release from the burdensome propinquity of the human species, quite unable to face them with equanimity. She skirted the square and the Old Town, and joined the river at the bridge on the Zenda road. From here she strolled through the dark, cool woods; the perpetual rhythm of the breeze-twisted leaves and the intermittent syncopation of bird song strangely soothed her unquiet mind.

She was invariably ashamed after such clashes of temperament, since she realised that both Papa Smith and Herr Josef wished only to help. She knew that she had behaved in a most ungrateful manner, but found herself unable to control the perversity that rose within her, sometimes without the slightest reason. She had not intended to be so lacking in manners; she had not intended to be angry. She had intended nothing. But something within her had been affronted, something within her had compelled her to demonstrate an outraged dignity which, though spontaneous, was quite without reason. She was becoming afraid of herself.

Papa Smith was quite lost in her seemingly uncontrollable changes of mood and saddened because of them. He felt, in his increasing inability to understand her, that his love for Isabella, and his eagerness and ambition to help her were as naught. Herr Josef, on the other hand, whose experience of the interpretive artiste was greater than Papa Smith's, merely waited, albeit with some anxiety, the final outcome of these wrestlings with the spirit. Always there was a struggle between the egotism of the artiste and the ordinariness of the human being. Sometimes the one triumphed; sometimes the other. It was rarely that both co-existed in a state of balance. Only time would tell.

Isabella could not abase herself so far as to offer an apology for her behaviour. Instead, on her return to her lessons, she approached her work with greater effort and a singular sincerity. Herr Josef accepted the situation without comment; watched over, guided and taught her. He gently tempered her hasty judgments on all things musical, checked her capriciousness, and accepted her arrogance.

When possible, and when appropriate, he praised her efforts.

The lessons went on. Isabella sang, and the quality of her voice increased in power and beauty. No reference was made to the recent disagreement, but always Josef felt that he was tiptoeing through a minefield. So perfectly were they once again able master and more than industrious pupil (at least to all appearances) that the misunderstanding might never have happened. In addition to teaching singing he discoursed gravely on the art of voice production, pointing his lessons with many anecdotes serious, humorous, even slightly scandalous, such as were listened to avidly by the matrons of Strelsau. From these Isabella also derived great joy, paying him the compliment of listening with respect and attention, and occasionally giggling at such innuendoes as she understood.

She called him *Maestro* again, in all innocence.

It was inevitable that during the practice of singing there were passages of tedium, for what is a glorious moment of perfection to an audience is the result of perhaps hours of wearisome repetition to the singer. No one is more quickly bored than the musical genius, of whatever species. But Josef had a dream to realise. If he noticed from time to time half-suppressed signs of recurring impatience, he ignored them. When hints of sophistication crept into her voice, he thought it meet to discourse on the phenomenon.

'Simplicity, Isabella, is the first rule of great art, especially of the art of singing,' he told her. 'Do not discover complexities of any form where there are none. Do not seek to put into the music more than is there already. If the composer be a man of genius, he has done all that is necessary. If he is not, we do not sing his music. Your part is to sing what *you* believe in, and to sing it sincerely. If you cannot believe in what you are to sing, do not sing it.'

Such advice was for the moment beyond Isabella's understanding. She grew quickly frustrated by the theme of 'simplicity'.

'I sing what I feel,' she retorted.

'So, my child, that is "simplicity".'

Nevertheless when she sang a song, or in particular, an aria, which was not to his liking, he would wag his finger at her over the forte-piano and say with a smile: 'Isabella, you are there interpreting the master's thoughts and you have no idea what they were. That is the critics' task; in you it is an impertinence. Let the music speak for itself.'

And even then, before she had well begun again, he could not resist saying, a little maliciously, perhaps to provoke her: 'You are not presenting either Fräulein Isabella or her virtuosity to the world. You are singing the music of Mozart, or Beethoven, as the case may be . . . .'

There was something to be said for Isabella's new-found attempt at self-control in that she did not burst out in anger: 'I hate you!' as she might once have done. Instead she smiled sweetly, and sang. It seemed to her that she did not know the sound of her own voice.

She sang with simplicity, and such was her gift that the music began to portray itself upon her face and her deportment. In her simplicity she did not play the character; she became the character.

In later years her audiences would exclaim: 'What an actress!' not knowing that what it considered consummate artistry was in fact the projection of Isabella Weber herself.

The battle was not yet won. The slightest gesture on her part, voluntary or involuntary, would cause Herr Weingartner to stop playing, to assume his benignest air — so like Herr Schubert, Isabella thought, with eyes smarting with suppressed tears — and repeat some odious advice.

'Dear Isabella, what the voice cannot express, the body should not. Indeed if the voice cannot express it, it is not worth expressing.'

His tone was mild, his eyes kind, but she could not mistake his meaning. She swallowed her pride — not without difficulty — and with equal difficulty set about ceasing to 'swim'.

As the lessons progressed other faults had to be corrected.

'Do not close your eyes when you sing,' said Herr Weingartner.

And when she opened them: 'Do not look up to Heaven.'

When she struck, by design or accident, what might have been a particularly meaningful attitude in the middle of an aria, he implored:

'Do not stand on tiptoe, either on one foot or two.'

Her eyes flashed, but her feet came flatly and firmly together, and remained so.

Day after day it went on. Herr Weingartner's salon became a battle ground of opposing temperaments whereon Isabella fought for the possession of her own soul: Simplicity. Simplicity . . . .

Do not . . . . Do not . . . .

Do not . . . . do notdonotdonot . . . .

Simplicity . . . simplicisincerity . . . . Do not donotdonot . . . .

At times Isabella was faint with frustration and anger. What should have been the easiest lessons to learn seemed to be the most difficult. Even when she was alone pride rose in her gorge.

'Do not. Do not,' she would almost snarl, her face contorted and ugly.

'Do not. Do not . . . . In God's name, is there anything I can do?'

Herr Weingartner smiled secretly. He confessed to himself that Isabella's recent docility had been a little trying, like living on the edge of a quiescent active volcano. It was too much out of character. Nevertheless there were signs that she was conquering her malaise of spirit very well indeed.

He looked at her over his spectacles.

'Yes, my child. You can sing.'

Isabella sang.

*

Herr Weingartner noted her efforts with approval, nevertheless, and with some satisfaction. Nor was he sparing in his praise.

'I think,' he told Fräulein Matthild' one evening after a particularly rewarding session, 'that we shall make something of our song bird after all.'

Matthild' laid down her needlework, and removed her spectacles with the deliberation that was characteristic of all her movements. For some moments of time she considered her brother, much as if he had said something she did not quite understand.

'Our song bird, as you call her, Josef, will eventually make something of herself, with or without your help.'

Josef stared at her, his pipe half-way to his lips. Matthild' frequently spoke her mind concerning the activities of her brother, but only when they were alone, and rarely in so forthright a manner as this. He was so taken aback that he did not even begin to bluster in protest, as was his habit sometimes when crossed.

Matthild' continued calmly, 'You know well enough, Josef, that already Isabella sings as well as, and in many respects better than some of the ladies who have charmed us with their artistry at the Opera in recent years.'

At the word 'ladies' Matthild's nose wrinkled slightly, and Josef laughed aloud at the inflection of the word 'artistry'. He savoured the primness and the precision of her pronouncement; with the truth of it he could do no other than agree.

'So much is true,' he agreed at last. 'And were it not for her age . . . .'

'Age, as measured in years, has little to do with it. You are blind, Josef. When she came to you she was a child, a pupil with a rare gift. You treated her as a child, and you have continued to do so. Furthermore, you drive her too hard.'

'So. She is ambitious. That is plainly to be seen. She has a divine gift, and wishes to use it.'

'Josef, Josef. How selfish you are.'

'Selfish! I?'

'Do not pretend, Josef. You are ambitious too, are you not? You picture her taking the world by storm with the beauty and perfection of her voice; you imagine her saying: "I was taught by the Herr Kapellmeister Josef Weingartner of Strelsau"; or perhaps you hear yourself proclaiming: "I taught Isabella to sing. I, Josef Weingartner." What folly, Josef. You will not make a singer that way. All that you will make is an automaton, a creature without life. At least all Herr Smith wished for was her happiness, but you . . . .'

Matthild''s voice ceased. In the silence of the evening room Josef felt bewildered, crushed. Never had Matthild' gone so far; never had she given so much as a hint of what she thought about his methods of dealing with the pert young Misses of Strelsau who came to him for lessons. It was not her business.

Matthild' spoke again, quietly, gently.

'You spoke of age, Josef. Have you not seen that Isabella is a woman, with a woman's thoughts and feelings? If she sometimes behaves like a child, have you not thought that it may possibly be your fault? She is in that shadowy area of indecision which is a torment to all human beings, be they geniuses or not. She needs sympathy, understanding, and gentleness. You give her work, discipline, monotony.'

'I give her music, Matthild'.'

'You give her boredom, Josef. She will, fortunately, be saved by her own obstinacy and impatience. I warn you, Josef, go carefully. There are many, many things about Isabella you do not understand. Nor ever will, I think.'

Miserably Josef admitted to himself that this was very probable. He made a pretence of fiddling with his pipe and tobacco for some minutes, seeing before him Isabella, standing straight and proud, hearing that faultless, perfect voice.

He lit his pipe, and with the rising smoke a weight seemed to be lifted from his mind.

'You are right, Matthild',' he said. 'I have been selfish in my thoughts about Isabella. Love of music must be my excuse.'

He puffed at his pipe.

'I will seek other advice. I must write a letter to Madame Marchesi.'

'Josef . . . .'

'Matthild'?'

'Already you have spent a great deal more than could by any consideration have been expected.'

'Matthild',' Josef said, earnestly, 'it is God's will that she should sing. Herr Smith has only the pittance the Opera pays him. What are we to do?'

Matthild' replaced her spectacles firmly on her nose, glanced once in his direction, and continued with her needlework.

The letter to Madame Marchesi took a great deal of composing, for in addition to the details of Isabella's natural gifts, her training by him, and her subsequent development, Josef considered that it would be as well to advise Madame of other personal characteristics she could expect to encounter which seemed to have little connection with the voice he was proposing to introduce to her. This was the most difficult part.

. . . Here is a singer for whom, even at this stage, no words of praise could be considered excessive or exaggerated. You will hear for yourself. But, gnädige

Frau, *she presents to me, to say the least, an intractable human problem. I have to confess that in dealing with this aspect of her development should it have fallen within my competence, I have failed lamentably. It needs, I am sure, the intuition, the understanding of another woman. You,* gnädige Frau, *will . . . .*

Poor Josef, unused to literary expression, could not find the words with which to explain his dilemma, and Madame Marchesi, for all her experience with budding female singers, was for a time, similarly nonplussed. With large flourishes she replied to him, after a month of working with Isabella in Vienna:

. . . I find her fascinating; at once a child and curiously adult. What factors have contributed to this state of affairs I can by no means understand. Your account of her early life leaves much to be desired, but it is undoubtedly that, and her development as a person, which is responsible for her present uncertainties, exaggerated, my dear Herr Weingartner, by your own unfortunate attitude towards her. However, I agree in general with your judgment of her talent. I regret that I find myself in disaccord with your method of developing it.

Josef sighed, and hid the letter in his cabinet.

*

Christopher Smith arrived in Strelsau late one evening after a tiresome journey across Europe, partly by Diligence, and partly by railway, the latest completed section of which reached the city via Dresden and Zenda. He was tired and hungry, but his fatigue and his hunger were both temporarily forgotten as he descended from his fiaker at his father's address, for from an open window on the third floor above him there came the sound of singing the like of which he had not, in his brief musical experience, heard before, the sound of a voice which not even the drab surroundings and the terrible echoes in the narrow cobbled street could diminish in glory. His baggage dumped in the street, he was left standing before the half-opened, undistinguished entrance gate, beyond which he perceived whitewashed walls and a colourful garden. The heat of the summer evening was lifting; dusk was falling, and the scent of the flowers was strong. In the street shadows appeared one by one; isolated snatches of murmured conversation were gradually to be heard as the shadowy figures drifted towards him, encircled him, inspecting him. Somewhat perturbed (for an English boy recently at school must attract some attention standing amidst a pile of baggage in a street in Middle Europe), he tried his schoolboy German on the nearest of the shadows.

'Who is it who sings?'

No one spoke. They continued to stare at him. He tried again; he felt himself blushing.

'Who is it who sings up there?'

He half expected, when they finally understood his barbaric pronunciation, to be given the unpronounceable name of some local *prima donna* not yet heard of beyond the frontiers of her own wildly beautiful but, to him, uncivilised country.

An old crone pushed herself forward, leaning on her stick. With a leer, she said: 'That is our Isabella.'

He thought this over for a while. 'Isabella? You said Isabella?' He pronounced the name very clearly, lest they should still not understand him.

'*Ja, ja*. Isabella.'

'Perhaps some famous singer,' he half asked, half stated.

The old crone cackled. '*Nein, nein*. Not a famous singer. Our Isabella. She is but a child; she lives with the Engländer up there.'

'One day she will be famous,' a second shadow offered an opinion.

'Hush,' protested a third. 'She nears the end.'

It was clear that they were accustomed to hearing this piece of Isabella's repertoire. Again they were silent, their faces now upturned, waiting. The sky became darker. The dark gables of the old houses stood out against the feeble light of the first stars. A few windows were suddenly bright with first-lit lamps. On the now gently-moving evening air, floated the last bars of the *Hoffnungsaria* with a perfection and a purity that was unearthly. Hope, rising with the stars, was triumphant over all the dark oppression of life, over the weak and timorous bestiality of man to man. Surely this was the voice that the Master Beethoven had written for, had striven to hear through his illness and cruel deafness.

'Our Isabella,' they said, with pride in their worn voices. 'She is but a child.'

It was not possible, Christopher told himself. It must be that they spoke of her with some local inflection of familiarity, with some pride in the achievement of a woman of mature years whom they had known as a child.

Isabella. Tired memories stirred in Christopher's mind. His father's letters, rambling sometimes to the point of incoherency . . . *I have adopted a little girl. From the poor area of the town* . . . the 'old man' had written, going on to describe the pity he had felt for her condition before digressing, as if she were of little importance, into some chit-chat about the foibles of orchestral players and the vagaries of conductors.

Christopher had paid little attention at the time. A boy at school has other things to think about than the meanderings of a far-away father. But now it all came back, and because time has variations in tempo all its own, that 'little girl' was to him still a little girl. His father had not said anything about singing, and Christopher did not connect the velvet voice he had just heard with a little girl.

Somewhere there was confusion.

The singing died away. There remained the silence of the velvety night and the emptiness of the cobbled street from which the shadows were beginning to melt away.

Suddenly a great bell began to toll the hour and the passing of the long day. Slowly, deliberately, it pontificated that Hope was human and temporary; that time only is eternal, and time knows neither Hope nor Despair.

*

The brazen notes fall ponderously upon the quiet city; the widening circles of sound splash into every seen and unseen corner of the streets; even, faintly, into the castle on the hill, whose Prince is sitting uncomfortably in the Royal Box, his eyes fixed upon that distant figure on the dim stage; it eddies into the salon of the former home of Herr Josef Sigismundus Weingartner, bereft many years ago of its singular owner and his sister Matthild' (he dead — a great loss to the city — of a heart condition; she late espoused to the Doctor who so carefully attended Isabella in her hour of darkness), now transformed and occupied by a rich manufacturer of chocolate patisserie; it seeps only faintly into the Opera House, where it is immediately extinguished by the great roar of applause which greets the end of the last aria of Act One.

Peccavi, one of the greatest of operatic arias, is finished. Backstage little groups stand silent, as indeed they must. Their hearts are too full for words. But in front the House is unable to contain itself, even though it is only the end of the first act. It rises to its feet as one; it claps, it cheers. Never has there been such a demonstration in this theatre. From all parts of the House they call her name possessively, insistently. They call for a repetition.

Christopher steals a glance at his watch; the piece is running several minutes late, even taking into account the late arrival of His Majesty at the theatre. Subconsciously he shakes his head at the insistent calls for an encore. There remain still two acts, neither containing anything more gruelling than the first, but still extremely demanding. An encore is unthinkable. Surely she will not . . . . But his is not the decision.

Nevertheless the applause continues unabated. In the wings they wait — Christopher, Maria and Hans (greatly privileged), both weeping copiously, the stage manager, the various mechanics, impressed in spite of their crude and practical natures.

Isabella is unique in the operatic world in that she has never encouraged *intemperate* applause, merited or not, nor even acknowledged it. Early in her public career she decided that she would never sing an encore. You do not demand of the baker, she is reported to have said, that if one of his loaves is good, to supply another similar one free. *I sing as well as I am able, she told an over-demanding audience. It is not necessary to spur me on to greater efforts. I am not a horse.* Adopting this attitude she was quite without the support of all other artistes in the musical world; they considered that she was quite mad, and did not hesitate to say so, for their egos thrived on praise and adulation, and jealousy was the driving force of their existence. Isabella ignored them, and remained adamant. She went further: when later, she was internationally famous, she would not enter into a contract with any Opera management unless the programme carried a note to the effect that Madame Weber considered that applause directed towards her personally was unnecessary, and she would be grateful if patrons would refrain from such demonstration. Should they really desire to show their appreciation further than by simply attending the representation, there was a collection box in the foyer for the receipt of Madame Weber's chosen charitable activity — the care of orphan children.

So shocked were her audiences by this that the orphans benefited greatly.

The applause continues, growing louder. Everyone waits for some sign of Madame Weber's displeasure at this unlovely demonstration, some searing flashing of anger, some humiliating rebuke. None comes. In the warm theatrical dusk she stands, a still white figure, her face a mask of ineffable sadness and beauty, the face of a mother contemplating her beloved importunate children.

Andelli, frozen upon his podium, seeks her eyes with anxious eyes. With the merest glance of acquiescence in his direction, she moves a little forward downstage. Andelli's mobile Italian face reflects a maelstrom of confusion — anxiety on account of Isabella, dismay and anger that the audience should so far have forgotten itself to behave thus intemperately and incontinently in opposition to her stated wishes, these are mirrored in turn. He bows briefly in her direction, turns to the expectant orchestra, and bites out the word 'repeat'.

Fearfully, a little behind her, *Marella* and the *Father Abbot* wait. To Lisa in her *debut*, and still unpractised in the usages of the stage, every second is an eternity. She seeks Madame Weber's eyes too, but vainly, for some sign that all is well.

Madame Weber looks beyond the cheering multitude of her worshippers to the Royal Box, where His Majesty Karl Ernst Maria Michael Frederik, now King of Ruritania, sits immovable, with a heart of stone.

The Ladies and Gentlemen-in-Waiting are silent too, and uncomfortable. They are aware of a kind of antagonism flowing between the Box and the Stage, a battle of wills between a reigning prince and a reigning prima donna. During an actual performance, such a situation is vastly intriguing. Several of these ladies and gentlemen were of the opinion, though without foundation of fact, that Madame Weber was singing *at* him when, on his coronation, she should have been singing *for* him.

Biting his lip in frustration, Christopher consults his watch again. It is now ten minutes over the time, and two acts to go. Nothing can be done. He will no doubt be acquainted in due course with the reason why Isabella, after a professional career of outstanding performances, has chosen to repeat, uniquely, this one, opera and aria; and why she has allowed such applause. Now, however, it begins to die away, and apart from a belated clap or two which echo about the vast arena, an expectant silence settles upon the audience. Green and white 'limes' present Isabella as a figure of ethereal loveliness. Not Isabella the actress; Isabella the voice made flesh. Proudly she stands, easily, motionless, her hands at her sides. Once again Christopher's heart overflows with love for her, that love which she had rejected but which would not be rejected, but continued to serve. He too, along with the Ladies and Gentlemen in the Royal Box, has perceived that Isabella is making an effect, is 'acting'. Quite unnecessarily 'acting'. Not, of course, in the crude and vulgar way of arm flinging and head turning favoured by lesser *histriones*, but in her well-considered way.

The words echo in her mind almost subconsciously. 'Sing only, my child,' says Herr Weingartner from his forte-piano. 'The work is greater than you. Let it take charge of you. That way lies true understanding and emotion.'

Dear Herr Weingartner. Involuntarily her eyes lingered for a moment on the stern face of Matthild' in the dim stage box, and turned away. Too many memories . . . too many. But now she will sing again, and her singing, if only for this one aria, will be spoken of with wonder and reverence for many years to come.

Andelli is of the same mind, for reasons he cannot put into word or thought. If there is to be a repeat, it shall be a repeat that the world will remember. He taps the edge of his desk; he raises his baton. On the downward stroke the compulsive beat of the 'cellos begins, followed by the roll of the double basses. On the violins the silver jet of a scale rise up to that notorious C sharp in Alt. From Isabella's incomparable throat the notes glide effortlessly.

Ah. Non posso far', Signori . . . .

*

The perspiration gathers on Christopher's forehead as it had as he stood in that narrow street in the slums of Strelsau. Even before he had laid eyes on her, he had fallen in love with her voice. He was suddenly aware that there was no further conversation, no sound except that of ghostly feet making dry little noises on the cobblestones, like the noise of dead leaves disturbed in the forest, as the shadows melted away, until he was left alone with his pile of baggage, wondering whether the voice he had just heard could really be as wonderful as it had seemed to be, or whether — because of the fatigue of his journey, and his hunger, both of which were now causing him some discomfort — his senses were playing him some trick.

The room with the open window on the third floor was suddenly illuminated by the soft glow of a lamp, and a figure appeared silhouetted in the golden rectangle.

A girl. A child, no more than a child. Christopher was aware of a vague sense of disappointment. His father had indeed properly written: *I have adopted a little girl, from the poor area of the town . . . .* But even in the darkness he could not help but be disturbed by the shadowy young face and the dark bronze hair haloed by the light within the room.

A little girl. A slum child. This could only be the child his father had written of. But the singing . . . . Surely there was a misunderstanding somewhere. Perhaps they had been teasing him, a foreigner, with an imperfect knowledge of the language.

The child was still at the window. In his father's apartment. She was gazing unmovingly into the street. The halo of hair was very attractive. Remembering some later references — no more than a sentence or two — in his father's letters, Christopher might reasonably have expected to be confronted by an earthy Bohemian face of the peasant type, with unattractive yellow hair pulled tightly back from the forehead and bound in the two pigtails which were such tearful features of Central European girlhood. He might have expected, a stocky, broad-hipped female creature with the potentialities of a good *hausfrau*. Perhaps that was why his father had taken the extreme step of 'adoption'; loneliness, and a growing boredom with such trivial household duties as could not well be ignored. In fact, for 'little girl', Christopher had understood 'servant', believing that girls began such lives at a tender age, in the more remote parts of Europe.

The figure at the window, then, supposing she were indeed the singer, could scarcely be the child his father had adopted, and yet, the name 'Isabella' . . . . There must surely be more than one Isabella in Strelsau. It was mere coincidence, after all. He began to hope that the child at the window might indeed be Isabella, the Isabella of his father's letter. How much more pleasant it would be to have a singer for an adopted sister than a pudding of a servant girl. Lost in the thoughts engendered by this idea, he did not realise that she was addressing him until she had repeated her words three times. When at last her voice penetrated his consciousness he drew himself up to a position of polite attention, thankful for the darkness which hid his schoolboy blush.

'I said "Good evening", three times.'

'I beg your pardon, Fräulein.'

'You were dreaming, boy, down there in the street. What were you dreaming about?'

Taken aback by the directness of this attack, he felt himself impelled to refute the charge.

'I was not dreaming. I was thinking.'

She laughed, and his foolish heart leapt within him.

'You were dreaming, boy,' she replied, this time, to his greater confusion, in English.

'I was thinking,' he repeated. Even to himself the phrase sounded foolish.

'Why were you dreaming, boy?'

He could not have explained to her what effect that beautiful voice had had upon him. He could not, for any reason, have explained that while he was listening to that exquisitely sensual singing he could have slain dragons. He would not have explained, in any case. He was English, a Public Schoolboy. One has one's reservations.

Now, momentarily, he felt inferior, at a disadvantage.

'Someone was singing,' he said. 'It was very nice.'

She laughed again, and his defences fell lower.

'That,' she replied, 'is truly English. What a wonderful word is "nice". It is good for everything. There is no word in German like it. What nice singing. What a nice evening. What a nice boy — ' she teased him, — 'standing dreaming in the street . . . . However do you know what you are talking about in England?'

Christopher perceived that perhaps there were serious defects in his education. Above him, leaning slightly from the window the better to inspect him, and sensing that he had been sufficiently mentally braced, Isabella changed her tone a little.

'It was I who was singing, boy,' she informed him. 'I sing quite frequently.'

'It was — beautiful,' he said. He had, quite without thinking, been about to repeat the word 'nice', but remembered in time.

And at that she laughed again, and cried: 'Bravo. "Beautiful" is much more complimentary than "nice". And who was it that clapped?'

'Clapped? No one clapped.'

'Someone clapped,' Isabella told him decisively.

'There was only myself here,' he looked around at the dark empty street.

'There was only myself and some — some others.' He could not bring himself to say peasants.

'You clapped, Christopher Smith,' she told him, and before he could recover from this further shock, she went on: 'Did you want me to sing again? It would have been useless. I had finished. I only sing when I feel like it. And I never sing the same thing twice. Not immediately, that is.'

She was stating a simple fact in perfect sincerity. Christopher at that moment considered her to be a little prig. He was making a mistake that many made in later years. But they have all had reason to change their views. Without exception Isabella has adhered to her own idea concerning applause and encores.

Except on the night of the coronation of His Majesty Karl Ernst Maria Michael Frederik.

On this occasion the audience, fully acquainted with her *fiat* on the matter of applause and encores, has chosen to ignore it, and she has chosen, of her own volition, to the astonishment of all, to break it.

Isabella continues the aria, and finishes once again on the sustained B Flat without showing the slightest discomposure. As the last vibration of the note dies away into silence the House erupts into a staccato crepitation of clapping mingled with a confused roaring and shouting. The noise echoes round and up into the great dome where cherubim play with lyres and ribbons. From fauteuils to gallery the movement of clapping hands resembles a swirl of excited butterflies. Even in the Royal Box, the Ladies and Gentlemen-in-Waiting throw discretion to the winds and clap like everyone else; it might be supposed, so enthusiastically, but genteelly, do they applaud, that they had paid for the privilege.

His Majesty remains silent, unmoving. His eyes are fixed on that distant shimmering figure as if he would annihilate time and space by the sheer effort of will, and draw her to him.

The overwhelming volume of appreciation sweeping over and about Her Majesty Amelia Viktoria Konstanzia Camilla has finally penetrated her phlegmatic lethargy and convinced her that something out of the ordinary must be happening. She surveys with glaucous eyes the mass of heaving patrons around and below her, the sweating musicians, and finally, Madame Weber herself, standing immobile before the footlights, bathed in phosphorescent green and white, her supporting artistes and the members of the chorus maintaining the frozen attitude by an extraordinary effort. Her Majesty can make nothing of it. Madame Weber is handsome enough, in a theatrical kind of way, she concedes; and sings well enough, she supposes. But since Her Majesty knows nothing of the art of singing, she is at a loss to understand what the vulgar demonstration is all about. Neither Her Majesty's ears nor her opera glass succeed in conveying to her the magic that makes men mad. She clenches her teeth in a catlike grin of non-comprehension, a delicate expression which is unfortunately disturbed by a yawn. She puts down her opera glass, and takes up another bonbon.

The applause continues.

It fills the huge House to overflowing, and seeps through the doors exquisitely padded with white leather and studded with purple bosses. Beyond the doors it is ingested by the sound of the great bell in the tower of the Cathedral of St Michael, forming a swirling pool of broken time which crushes life and fame and applause beneath its unceasing, unrelenting ebb and flow.

This fearful destruction of Time cannot fortunately be heard inside the House; nor can the tinkle of the little travelling clock as it plays on the quarter-hour its *perpetuum mobile* to the empty dressing room. Bells and clocks are the accidentals in the symphony of time, not to be taken too seriously.

But still they clap and vociferate. Still they call her name and shout *bis* repeatedly. Christopher consults his watch again, with a frown. It is indeed the quarter-hour. The clapping seems to be a little less spontaneous; it seems to have lost the edge of sincerity. The ovation is becoming hysteria.

'Isabella!' they call, the familiarity supported by anonymity. The name certainly trips off the tongue in easy syncopation. The sound of it dances away over the staccato clapping; it echoes from wall to wall, over and across and spiralling ever upwards so that its vibrations dislodge the slight film of dust from the pink cherubim and their ribbons and swags. The dust dances in the beams of the limelight.

'Isabella!' they chant.

'Isabella! Isabella Weber! Isabella Weber! Isabella Weber!'

'Isabella! Isabellabella! Isabellabella Weber!'

'Isaisabellabella Weberebererber! . . .'

Isabella makes no movement. Her face has become cold, stony. Christopher understands the contempt with which she must be hearing this . . . patronage. The word is an obscenity.

In the green and white light, Isabella is seen to put a hand to her breast, a slight histrionic gesture perhaps, in acknowledgement of the still-roaring audience.

Maria's scream is heard above the noise.

'The curtain! The curtain!'

Almost before she has stopped, the curtains come together, to the consternation of the audience suddenly bereft of its idol. Christopher, darting across the stage, catches Isabella on the verge of collapse, lifts her, and carries her off to the dressing room.

The mechanics stand respectfully aside, shocked by the turn of events, until they are roused to action by the lashing tongue of the stage manager, who demands to know whether, in the first place, they wish to give themselves the pleasure of earning their miserable wages by condescending to change the set, or in the second place, whether they wish to return to their homes this evening. The complete effect of this verbal excoriation is unfortunately lost, however, since they move to their allotted tasks as soon as he opens his mouth. Under their seemingly disorganised activity rocks melt and disappear; mountains withdraw into the purple distance; the walls of the Church of The Holy Sepulchre fly apart at the hands of those whose lives are spent behind the scenes, and come together again in the shape of the Inn of the Cat-in-Boots; the Churchyard becomes the village square; flowers grow where previously there were sharp religious stones; in place of the evening stars there is — momentarily flickering — the midday sun.

The stage is dressed for the second act, and now waits, deserted and gloomy. Before the curtain the lights are bright in the seething auditorium. The Royal Box is empty, discreetly and feudally veiled. The royal couple, their respective Ladies and Gentlemen-in-Waiting, have withdrawn for the period of the intermission into the salon. The audience, released from the discipline of protocol, moves also. It swirls, coalesces, turns and twists in the direction of the promenade, where it will discuss at length, repeatedly and with animation, the most brilliant singing that has ever been heard in this or any other opera house.

Madame Weber's collecting box for orphan children is soon filled to overflowing, and the consciences of the good burgesses of Strelsau are eased accordingly.

Madame Weber lies unmoving on the chaise-longue where Christopher has placed her. Her face is pallid beneath her sparing make-up, and her forehead moist with a greater amount of perspiration than the effort demanded by the performance would warrant. The doctor, bending over her with his stethoscope shows grave concern. The perfume of flowers is overpowering in the room; they are too many, they have the scent of transfiguration and apotheosis.

The ghosts of times past shimmer in the looking glass, and hide gibbering and chuckling behind its dark panes.

'Why did you clap, you and the others?' some mouth in caricature.

'It would be useless,' others reply, 'I have finished.'

'Why did you clap?'

'Why did you clap, boy?'

'Stop!' cries Christopher violently. 'I did not clap . . . .'

The doctor straightens his bent back.

'*Mein Herr*?'

'I am sorry, I was dreaming. I blame myself. I should not have let it come to this.'

The doctor favours Christopher with a searching stare.

'Perhaps you are right. However . . . .'

His eyes are blue and disconcerting. He has witnessed the development of many dramatic situations behind the scenes. Older and wiser now, he is no longer moved by the little troubles of the great.

'I can scarcely imagine that you would have reason to blame yourself, Herr Smith. The disposition of humanity's ills is not within your competence. That matter is in the hands of God.'

'But Isabella . . . Madame Weber,' cries Christopher.

Her face is still grey; her make-up now an unlovely smear under the rivulets of perspiration.

'Perhaps,' the doctor continues, 'there is no great cause for apprehension. Perhaps there is every cause. Medicine is far from being an exact science. The truth of the matter is that I find a slight irregularity in the action of the heart.

You and I both know, Herr Smith, that this is a long-standing condition, do we not?'

He gives a discreet medical cough.

'With due consideration,' he struggles to give the right importance and value to his pronouncement, 'with all due consideration for what is at stake, and in view of the occasion,' again the little cough and the innocent stare of the blue eyes, 'I must advise that Madame Weber withdraws from this — er — performance — er — presentation, a little rest, and then . . . .'

'It is scarcely possible,' Christopher interrupts. 'We have no understudy. All these good people are under contract. His Majesty . . . .'

'Aha,' says the doctor. 'His Majesty. So! May I recall to you that the preservation of life is greater than the pleasure of kings? I forget who said it, but it is none the less a continuing truth.'

Footsteps echo agitatedly in the corridor outside; there is a knock, followed by the turning of the handle. Maria's face is grim, belligerent. She strides purposefully to the door, a flask of *sal volatile* in one hand and a fine lawn diaper in the other.

'I will have no others in this room,' she pronounces. 'Already it is too full.'

But Andelli is a more than privileged person, and it is obvious that he has great reserves of recuperation. Already he has replaced such garments as were soaked with perspiration, and in everything except his wild mane of dark hair he is point-precise. He exudes a beautiful perfume, more natural than that of the flowers; or so the label on the bottle tells him.

He flings himself across the room, his preposterous personality dwarfing the others.

The entry of this *Furioso* causes Isabella to open her eyes. She essays a weak smile in his direction.

'Andelli, as always,' she murmurs.

'Madame,' he cries. 'Wonderful. Superb. Which?' he demands with a gesture that takes in mortals and immortals alike. 'Which is Cordelia? Which Isabella?'

'Mountebank,' Christopher says audibly.

Andelli scorns him as only an Italian can.

'Madame, Isabella, you have never sung better. Your voice is a miracle of sound. But why, in the name of Heaven, did you repeat? We are at the end of the first act only, and we are late. Yet you continue to sing like a nightingale.'

Maria takes Andelli by the arm and propels him towards the door, thrusts her broad peasant face into his.

'Can you not see, *Signor Capo di Orchestra Andelli*, that even a nightingale may sing too much? Out with you.'

Andelli is aghast. He has not been so addressed since he was an unshaven student. Indignantly he draws himself up to his full height. To the top of his hair he is a full centimetre taller than Maria.

'*Signorina Servente*,' he hisses icily, 'you go too far.' Taking a firmer grip on his arm, she replies in his own tongue, with a rather vulgar intonation.

'Not as far as you are going, Maestro. Out with you. All of you, out on your own feet before I take my broom and sweep you out. Come, Doctor, you too. You can do no good here for the moment.'

'Isabella,' Christopher begins, appealing to her over Maria's broad shoulder.

There is a ghost of a smile on Isabella's lips.

'Go now,' she whispers. 'Go and dream awhile, Boy. Have no fear, I shall finish the representation.'

Maria opens the door so violently, thrusting Andelli before her, that the two of them all but stumble over an embarrassed flunkey with raised arm in the act of knocking. She pushes Andelli into the corridor, and invites Christopher, the doctor, and Lisa to follow him. Standing astride in the doorway, a menacing fury with arms akimbo, staring out of countenance the beardless egg-faced flunkey, she invites him to state his business as briefly as possible.

'His Majesty . . . His Majesty . . .' stammers the youth, thrown into a nervous agitation by the strong resemblance of the figure before him to that of his maternal grandmother.

'What of His Majesty, dolt?'

Maria continues to stare. Her mouth hardens; her eyes narrow.

'Continue, Signor Carpet Knight,' she encourages him.

The youth swallows.

'His Majesty,' he parrotted, 'begs to be allowed the favour . . . the favour of presenting himself..himself to Madame Weber . . . as soon as . . .' His voice died to a whisper.

Maria releases her compressed lips, and permits herself a crocodile smile which is quite alien to her nature.

'So.' She observes, when she has digested the full purport of this message.

'So. His Majesty *begs*, does he? A tame dog, by the bones of St. Joseph. He was not so given to begging once, when his blood was hot and his desires lusty. Return this message to His Majesty, Boy, with all speed and propriety. Tell him: *Madame Weber begs to be excused for not receiving His Majesty*.'

'But, Madame,' the flunkey begins, aghast at such impertinence. He recovers something of his courage. 'In Strelsau His Majesty's wish, however expressed, is another's command.'

'You may add, should you deem it pertinent,' Maria allows the addition of the words '*at this point in the drama*.' 'That may soften the blow somewhat, and save your neck. What may happen hereafter is the will of God.'

Knowingly she repeated the doctor's phrase.

'Madam,' pleaded the flunkey, now plainly terrified, 'such an answer is not expected, I assure you.'

Maria's voice has a keen edge.

'What is not expected is invariably inevitable. You have your answer, is it not understood? Madame Weber receives no-one. But no-one. Away to your master, and acquaint him with that fact.'

The candles flanking the looking glass tremble in their sockets as she closes the door with some violence in his face. The light of their dancing flames breaks into a thousand pieces, sliding and shuddering across the walls and ceiling. Peering back through the looking glass, obscured by the dancing shadows reflected in its silver depths, the ghosts shriek with merriment.

'We told you so,' they mouth, falling about themselves. 'We told you so.'

Maria, shocked now by her assumption of authority, falls to her knees before Isabella.

'Madame,' she pleads, 'your pardon.'

'It matters not, Maria. You thought my thoughts; you spoke with my tongue. Now bring me my tablets and my smelling bottle.'

The little clock on the dressing table, as though to recall attention to the reality of the situation, begins to recite fussily that part of its repertoire that indicates the half-hour.

With a sigh Isabella relaxes.

'So much,' she murmurs, 'so very much, in such a short time.'

End of Act I

INTERMISSION

INTERMISSION

The snow was falling steadily on the cold leaden river; it was falling in the narrow winding streets and on the pointed roofs of the toy-like houses and in the square, muffling the diurnal noises of the city in a white blanket of silence, transforming it into a postcard scene: *Strelsau in Winter* . . . . Strelsau was indeed noted for this picturesque attraction throughout a continent which boasted many such attractions. Travellers came from far and near to revel in this materialisation of the illustrations of the books of their childhood. The City of Strelsau, they considered, might have been imagined by Herr Hans Christian Andersen in one of his recent stories. They soon discovered that the greater part of the city lay undoubtedly in the domain of the Brothers Grimm.

The exploration of the city's attractions was of necessity confined to the daylight hours, which at this time of the year were few in number. Although it was only late afternoon, darkness had already descended on the forests and mountains. In the narrow streets and alleyways oil lamps flickered at the corners and under the arches. The light they gave was minimal, as was the meagre illumination in some few of the house windows. Only at one house in the Square was there a little more light than usual; a large lanthorn above the door showed a little way through the snowy darkness, complementing the light from within. Herr Josef Weingartner waited the arrival of the doctor. He tramped to and fro before the half open door, beating his hands together or blowing upon his fingers like an outsize Pan upon his pipes, in an effort to drive off the cold. The snow gathered upon his head and upon his damp shoulders. He had no thought to remove it.

Peering through the curtain of snow he perceived a voluminously cloaked figure making its way towards him on uncertain feet. He hastened forward, a testy greeting on his lips. Taking the doctor's arm, he urged him towards the lighted doorway.

'Steady, man, steady,' remonstrated the doctor. 'I too am well nigh exhausted, trudging through this confounded snow.'

'But hurry . . . .'

'There is no necessity to hurry, my friend, more than we can. The patient is either quick or dead. The first, I presume, since you are standing here inviting your quietus also.'

The doctor stopped with his foot on the lowest step, and gave Herr Josef a quizzical inspection through the thick pebbles of his spectacles.

'You omitted, characteristically, and in your haste, to inform me of the identity of the patient I am attending,' he continued. 'You appear to be remarkably healthy still. Perhaps you suffer from a little over-indulgence of red wine, since your nose is confoundedly red. Or perhaps you have a simple rheum, since I observe that your eyes water and your teeth chatter like castanets. It is very fortunate that your bowels appear to be in good shape, as is your stomach.'

Chuckling at his own ponderous good humour the doctor clapped his friend on the shoulder.

'Confound your jokes,' Josef replied testily, for he was indeed cold and more than a little apprehensive. 'I am quite well. That you know. I myself have little use for abracadabra of your sort . . . .'

'But where others are concerned, we doctors are a necessity, eh? Come, man, who then is the patient? Perhaps our good Matthild', repining because of some imaginary spinster's flux?'

'Matthild' is well enough also, for aught I know,' Josef told him, shortly. 'The matter is more serious. My pupil Isabella Weber has been suddenly taken ill, and that fact is of some importance.'

'Our little nightingale, eh?' The doctor pursed his lips, suddenly very solemn and serious. 'That is indeed a different affair altogether.'

Within the house the doctor removed his outdoor garments carefully and with precision, while Herr Josef recounted how Isabella had been singing, quite happily, he insisted, and Matthild' had been engaged upon her needlework, altogether unmoved by the fact that she had been listening to a future *prima donna*.

'Yes, yes,' nodded the doctor in agreement, 'and then?'

He, Herr Josef, had at that moment been occupied with a complicated piece of forte-piano redaction from *Don Giovanni*, and knew nothing of what had happened except that Isabella had ceased to sing. Matthild' stated that the child had appeared to be suddenly distressed; that she had put her hand to her breast; that she had sunk to the floor from where, after the shortest possible moment of indecision he, Josef, had lifted her delicately and laid her upon the chaise-longue. Thereafter he had inefficiently endeavoured to introduce a little cognac between her blue lips and clenched teeth, while Matthild' waved the smelling bottle before her dilated nostrils. These combined efforts to restore Isabella to consciousness by means of homely remedies proving of no avail, Josef had bethought himself that the doctor would have more success. To this end, he had despatched a passing urchin, with a monetary stimulus, to run for that person. It was imperative . . . .

'Yes, yes,' said the Doctor again. He was now perspiring in his turn, for Matthild', desperately believing that everyone and everything must be kept as warm as possible, had put a great quantity of logs into the stove. They were beginning to burn furiously. The stove was a miniature furnace.

Isabella lay on the chaise-longue, a still, white crumpled figure, bearing little resemblance to the proud singing bird Herr Weingartner had tended, as he thought, so carefully, and for whom he had secretly made such wonderful plans. There were tears on his cheeks as he watched Matthild''s still ineffectual ministrations with the smelling bottle. She was an indifferent nurse, but she did all she could think to do. She applied the smelling bottle; she chafed the cold hands; she moistened the lips; she carefully removed the cold perspiration from the pallid brow. When she rose from her stool beside the unconscious child to greet the Doctor he saw that there were tears in her eyes also. How strange, he thought, that the sickness of a child unrelated to them, could move two such self-sufficient people to tears. Never, in all the years he had known them as medical adviser and personal friend, had they disturbed their own comfort for so much as a sparrow — if an exception were to be made of the cage of singing birds, now covered with a snow-white cloth, lest their untimely discord should disturb the invalid.

'She is my best pupil,' Herr Weingartner mourned in his distress. 'Never has anyone sung as she sings. No one every will . . . .'

The tears coursed unchecked over his rotund cheeks, and his nose grew redder with an emotion he could neither express nor repress. The long hours of practice together ran in his mind; recollection of the childish arrogance that had temporarily frozen his heart, and the humility and sweetness that had melted it again. He recalled his untimely impatience with her witless faults, and the arguments they had indulged in, she the innocent, he the fatuous sophist.

He agonized at the enormity of his dereliction.

'It is my fault,' he protested. 'I have made her work too much and too long. It is my fault.'

His sister showed for a moment some of her habitual asperity.

'No, Josef,' she contradicted him. 'It is not so. She has never done more than she was able. And no girl ever fainted from "hard work". Females in general faint from lack of work, and perhaps related causes.'

This last phrase she voiced almost in a whisper. The Doctor caught its delicate innuendo, and smiled briefly.

'Nothing of that kind of thing in this case,' he said, without specifying which of the two.

'What other cause could Isabella have?' cried Josef.

Matthild' sniffed loudly, but her lined face became gentle again, and she made a gesture she had not made for a number of years. She moved to his side and put an arm round his shoulder. Together they stood, like two bewildered children, their eyes upon the Doctor's face. On one knee beside the supine figure he noted with some interest this tardy expression of affection and support. But even before he applied his stethoscope he was of the opinion that here was a more serious matter than a mere attack of the vapours. His face was grave as he rose to his feet.

'You said she was singing?'

'That is so.'

'What was she singing?'

'What was she singing?' Josef repeated. He was completely at a loss. One's favourite pupil was taken ill in one's home; one requested the presence of a medical friend; one's medical friend preferred to discuss his patient's repertoire rather than proceed to a cure. If one's doctor were not also one's friend . . . .

'Monstrous,' spluttered Josef, 'monstrous . . . .'

The Doctor was patient. 'You are overwrought, Josef,' he said. 'You know full well, should you pause to consider it, that I have my reasons for asking such a question. And if you feel disinclined to answer it, I will satisfy myself.'

He trotted to the forte-piano and took up the book which stood open on the stand.

'This, I suppose, is what she was singing?'

He turned the pages of the work, seeking the title. *Don Giovanni*. He scanned the score, noting the runs, the *arpeggi*, the intervals. Finally he read, slowly, the text appropriate to the music, and nodded his head once or twice.

'There is little that should cause physical trouble there,' he decided, 'other things being in order. Our Viennese friend Jung might have some ideas on the subject. Very interesting — prescience, or nescience? I wonder.'

He turned suddenly to Josef, who was bemused by this recondite obscurantism. 'How long were you and she rehearsing this peculiar work?'

'Perhaps an hour,' Josef replied miserably.

'Perhaps an hour. H'm. Was there any complaint of sickness during this time, any vertigo or faintness?'

'She had complained of a slight headache.'

'So . . . .' The Doctor rubbed his hands together.

'And would you know the child's age?'

'Isabella is sixteen years of age.' It was Matthild' who answered.

'Now, friend Josef, I wish to make a complete examination, and I naturally require your absence. Matthild' will stay and assist me.'

Reluctantly Herr Josef left the room. As the Doctor went about his examination of Isabella, Matthild' questioned him from time to time with her eyes. He gave no sign either by word or look that he had discovered anything untoward. When he had finished, it was Matthild' who was perspiring, partly perhaps from the heat of the room, but chiefly from nervous tension. When at last he spoke, his voice was gentle.

'It would be unwise to replace the clothing. Bed and blankets would be much more to the point.'

Pausing only long enough to take in the full purport of this statement, Matthild' left the room. The Doctor replaced his instruments in his bag, cleansed his hands with a kerchief soaked in toilet water, gazing the while with some compassion on the still unconscious form of Isabella.

'These perfectionists,' he regretted. 'They drive themselves — and others — to the devil.'

Matthild' returned carrying a substantial blanket large enough to contain a number of Isabellas. This she wrapped about the girl with a care and gentleness which won the Doctor's approval. He nodded, and smiled a little, gravely.

'Now,' he said, 'I think we may admit our good Josef once more. He has been alone with his thoughts far too long.'

It appeared, however, that Josef had not been alone. When he was readmitted to the salon he was accompanied by Herr Smith and his son to whom he had sent a similar message to that which had summoned the Doctor.

Somewhat fearfully they awaited the Doctor's report, none daring to be the first to ask lest he should receive the answer he half expected. The Doctor did not immediately enlighten them. Addressing Josef, he said: 'We are holding a reception, I perceive. I do not know these — gentlemen.'

Josef hastened to repair the omission.

'This is Herr Smith. The Englishman who adopted Isabella as a child. He is Leader of the Orchestra at the Opera.'

'So,' said the Doctor. He clicked his heels and bowed like a sparrow picking up crumbs. As he surveyed the Leader of the Orchestra, noting his lined face, his prominently-veined nervous hands, his almost threadbare clothing, his nostrils were wide with the distaste of the superior caste for the inferior. He could imagine no relationship, physical or spiritual, that could bind the blossoming aristocratic beauty of the unconscious child and this work-a-day musician.

'And this young man,' Josef indicated Christopher standing mumchance behind Herr Smith, 'is his son. He too is a musician. A violinist.'

The Doctor rejected this information as being unlikely to help in the situation, although at first glance there was something attractive about the tall youth hovering between boyhood and manhood with the face and hands of an artist, withdrawn, inward-looking.

The easy familiarity which the Doctor had used towards Josef and Matthild' had disappeared in the presence of Herr Smith and Christopher.

Consequently they were unprepared by any personal softness of manner for the shock of his professional pronouncement.

'It will be obvious,' he said, 'that for some time I shall be unable to make a final diagnosis. What I have to say now is in the nature of a personal opinion, based upon what I have been told about the patient, and what I see with my own eyes.'

He turned his deep-set eyes under their bushy brows upon Josef and Herr Smith in turn. They conveyed an unspoken accusation. But his words, when they were spoken, were precise and deliberate.

'The child lives, and will live. So much is clear to me. But I find a considerable depression of the heart, and irregularity of the pulse. The reason for this is not clear.'

He turned his eyes upon Herr Smith, who made a gesture of non-comprehension.

'Further information of the patient's medical history is essential.'

Herr Smith moistened his dry lips.

'There is no such history,' he said. 'She has never been ill. She has had good food, and always she has grown stronger and taller.'

The Doctor conceded the statement graciously.

'The patient appears to be properly nourished. I have no doubt of your care in that respect. As for the growing taller, h'm. The cardiac weakness is scarcely a matter of either quantity or quality of food. It is a matter of inheritance, exacerbated by conditions we have little knowledge of.'

He turned to Josef.

'You spoke of her as an orphan?'

'I said she was adopted by Herr Smith. I have no other knowledge.'

The Doctor turned to Herr Smith.

'She was a child of the slums; I took pity on her.'

The Doctor waited.

'She had an interesting voice, even as a child. There is nothing more I can tell you,' Herr Smith told him. 'She has been treated as my own daughter. She has had everything I could give her.'

'Her parentage?' the Doctor persisted.

Herr Smith remained silent.

'Come, man,' the Doctor persisted.

Herr Smith drew himself up and stared the Doctor in the face.

'Her parentage — is not pertinent to her treatment,' he told the Doctor, bravely. 'Certainly she was born into the best of conditions . . . .'

But there was little resemblance now to the child who had sold matches in the square, shivering in the depths of winter in her thin inadequate clothing, almost too cold and too tired to walk more than a few painful steps to ease the ache of her numbed feet. The picture, considered dispassionately in this comfortable salon was so fantastic that he doubted if it had been real. One forgets so soon the unhappy times of life. His Isabella was tall, strong, beautiful. Now, suddenly, she was ill. From what cause?

Had he driven her too hard? Had she worked herself too hard, driven by some inner compulsion, some purpose beyond the understanding of her years, some Daemon that would not let her rest in her search for perfection in her art?

Or was she impelled by some ancestral procreative urge in her search for an identity?

The Doctor addressed Josef once more, realising that nothing more was to be gained from Herr Smith.

'In addition,' he elaborated, 'there is a further complication of a nervous collapse involving a temporary paralysis of the motor centres. The severity of this we shall not know until the patient regains consciousness. The cause, you will ask — overwork, exhaustion, protracted strain, who can tell? The treatment, the only treatment possible at this stage, is complete rest, seclusion, absolute quiet. If this routine is not adhered to, I will not be answerable for the consequences.'

Herr Weingartner made a gesture of helplessness.

'How long,' Papa Smith almost begged, 'how long will she remain in this condition?'

'I cannot answer your question precisely. Some hours, days; some weeks even. If we are unfortunate . . . .'

He allowed them to supply the remainder of the sentence according to their own dispositions.

'She is not in pain? She is not suffering?'

'Herr Smith,' the Doctor was strangely gentle with him. 'Only she, and God knows how much she has already suffered, and how much she has still to suffer. That part of it is over for the time being.'

Josef Weingartner appeared to have aged considerably. He croaked the question the answer to which, after the question of life and death, lay nearest their hearts.

'Her . . . voice?'

'Josef,' the Doctor reproved him, 'I used to think sometimes that you had a metronome for heart. Now I know it. She will be fortunate if she loses only her voice. For myself, if I may make it clear to you, I find it remarkable that she still lives. I had not wanted to say it.'

For some minutes only the crackling of logs in the stove broke the silence that possessed the room and its occupants. At length the Doctor prepared to make his departure; he assumed his voluminous cloak and took up his bag. At the door he paused.

'There remains,' he said, 'to put the child to bed as quickly as possible. She should not be moved thereafter until I am satisfied that no harm will accrue from such an action. She must have constant care and attention.'

Matthild' gathered their bewildered thoughts into a decision of her own.

'She shall stay here. We have a comfortable room, and I doubt that Herr Smith is sufficiently accommodated in that respect. It would not even be proper, I think.'

Herr Smith was too shocked to do other than accept the situation, and Josef, seating himself heavily in his chair, could only stare at the cold white face, the closed eyes, the still blue lips as if he would restore them to their former animation by a mere effort of will.

'My poor nightingale,' he said. 'And no one has yet heard you sing.'

'My friend,' the Doctor replied with some assurance, 'God has heard her sing. I wish that you may all sleep with a clear conscience.'

The Doctor was a careful man; a precise man, a man not given to expressing his emotions where his craft was concerned. In his time he had seen much of the sufferings, real or imagined, of those who sang. He knew only too well how, behind even the pretence of suffering there sometimes lurked a great fear, the fear of the dedicated artist of losing the voice, the fear of loss of memory, the fear of losing an audience, or even a single patron. Always there was the prime agony of making the first entrance; the jealousy of other singers who know that they can sing better; the constant censoriousness of the critics, who know how it could be sung better.

And here were two grown men and a woman (*pace* Matthild'), to say nothing of a boy still wet behind the ears, proposing to toss a child into this nightmarish jungle, without the understanding of or the ability to maintain their valuable music machine in proper working order.

'Truly,' he reflected as he battled with the wind and the snow, 'musicians are, in general, a ridiculous incompetent affected species of creature. How is it possible for one to consider them in the least normal?'

*

ACT II, Scene i

The market place in the village of Buonasera. The afternoon of the next day.

STRELSAU — VIENNA — STRELSAU

ACT II, Scene i: The market place in the village of Buonasera. The afternoon of the next day.

STRELSAU — VIENNA — STRELSAU

The bitterness of the Doctor's reflections translates itself as he leaves Isabella's dressing room into the banging of the door, a gesture he does not normally allow himself, for any reason. At the same instant a particularly strident bell announces the approaching end of the intermission and the imminence of the beginning of Act Two. The unlovely sound reverberates in the empty narrowness of the whitewashed corridor, empty, that is, except for Christopher, who turns hopefully as the Doctor approaches.

'How is Madame?'

The Doctor considers him with some distaste.

'There is some cause for apprehension,' he informs Christopher. 'I have nothing to add to my previous advice.'

'We . . . . They are waiting. Nothing has been announced. Some sort of rumour seems to have got about.'

The Doctor shrugs his shoulders. A professional life spent as Physician to the Opera has left him somewhat apathetic to the conceit of musicians and those who have to do with music.

'I repeat, in my opinion it would be folly to continue. But I cannot compel . . . .'

'We must continue,' Christopher interrupts him. 'We have no other course open to us.'

'On your own head be it. *Glück*, my friend.'

As he makes his way back to his *loge*, where Matthild' awaits him with her usual imperturbability, he is made very aware that, to quote Christopher's words, 'some sort of rumour' has got about. Like the green bay tree it has spread from person to person, from *loge* to *loge*, from the auditorium to the corridors, and from there to the foyer, carried by the universal method of knowing looks, uninformed chatter and innuendo. No one knows, of course, what has happened, or if indeed anything has happened at all. They are at liberty to make the best, or worst, of what bits of conversation they can snatch out of the passing air.

Madame Weber is suffering from every conceivable malaise from migraine to total collapse. Madame Weber will not continue to sing; Madame Weber will continue to sing. Madame Weber begs the indulgence of her patrons (how they preen themselves!) and requests them to accept her understudy . . . Madame Weber . . . .

But who is Madame's understudy? That girl who sings the maid? A thin-faced chit of a girl . . . She sings prettily enough, in a clear drawing-room voice, not a voice for the theatre.

With long lower lips the audience does not anticipate, in view of the many rumours, anything outstanding in the way of musical interpretation for the rest of the evening. Nevertheless they must see it through, as a matter of duty.

Returning to their seats, their tongues still busy with speculation, they cast curious glances at the still-closed motionless curtains of the Royal Box. They are not to know that His Majesty has spent the long minutes of the intermission pacing with slow and thoughtful step, alone, along the soft-carpeted corridor, his head sunk upon his breast, looking neither to right nor left.

A bell has rung.

The corridor is peopled with sycophantic bows and curtseys which mark his progress as bending flowers mark the passing of a breeze. They are intrigued by the desperate aspect of his face, by the curve of despair which bows his shoulders. They exchange significant glances as they acknowledge the presence of majesty in their democratic midst.

His Majesty Karl Ernst Maria Michael Frederik stops; he lifts his head. On the instant the House Manager, resplendent in evening dress, the blue riband of the Order of St. Michael across his chest, materialises from his lurking place mincing like a pigeon in love. He claps his white-gloved hands. The white and gold doors open.

His Majesty enters his *ante-loge*, and there is immediately a perfumed agitation as disarray is made discreetly presentable. The exchange of excited chatter has given way to silence. The Gentlemen-in-Waiting assume the superior faces that are the result of daily association with royalty; they open the farther doors; they withdraw the curtains to admit the ebullience of conversation from below and around. Bare shoulders gleam, jewels flash, dresses billow and cascade, weave and interweave into a kaleidoscopic mutability of colour ordained by the mere lottery of a number on a ticket.

The soul of the theatre is returning to its body.

The Ladies-in-Waiting range themselves behind Her Majesty Amelia Viktoria Konstanzia Camilla ('Camilla' for her somewhat roguish Spanish grandmother, whose qualities have by no means rubbed off on her) as she rises to accompany her Lord and Master into the Royal Box. Her Majesty sweeps rather than walks, for she appears to have mastered the art of walking without appearing to walk. 'The movement of the limbs is so vulgar, so indelicate,' her childhood governess had said, (meaning perhaps: 'in your case'), binding her arms immovably to her sides, and hobbling her legs above the knees. One might suppose, as she progresses, that Her Majesty is a peculiarly unattractive waxwork propelled by some ingenious engine concealed beneath her skirts.

The Ladies-in-Waiting walk with suppressed giggles and delicately pointed tread.

The white and gold doors close silently behind them.

Other doors, plain green-painted doors open in the whitewashed corridor where Christopher stands fighting his anxiety alone. From these doors issue chattering figures in every kind of costume and make-up. They linger, they loiter, they gesticulate as they chatter as if performing in an opera were the last thing in their minds.

The Stage Manager deploys his forces with cunning strategy. He is the Officer Commanding. The plan of the backstage is engraved upon his heart. He knows every corner, every pillar, the disposition of every bank of scenery behind which a couple may linger unseen for a last touch of the hands and a last whispered word. When he has dislodged them from these shadows they will still find opportunity to touch hands again, even 'on stage', but innocently, as if it were all part of the action. That is beyond his control; he is now chiefly concerned to get them into their rehearsed and ordained places. He pleads; he perspires.

'On stage, please.'

He cajoles. He pleads again.

'Ladies and Gentlemen of the Chorus, on stage, please . . . .'

'Ladies . . . please!'

They run up steps, they run down steps, giggling and chattering. They converge upon him from all directions. The bolder of the young ladies pat his hand in passing; one or two bestow a passing kiss to add to his confusion, the touch of a butterfly wing, no more. They giggle as they pass.

His clothing is awry; his face is carmine. It happens at every performance. With a gesture of despair he follows his flock.

Another voice takes up a different refrain. To Christopher's ears it is an ominous refrain, foreshadowing a time of fear and foreboding.

'Principals for Act Two, please.'

The voice comes nearer, spiralling down the iron staircase, repeating its doleful cry. To Christopher it is the voice of dilemma.

'Madame Weber, principals for Act Two, please.'

Christopher's heart is in his mouth. But when Isabella comes from the room she is calm and dignified, completely in possession of herself, strengthened by some power of the mind which is hers alone. There is little sign of the recent indisposition; and if her eyes are brighter and more luminous, such brightness may be easily explained by the contagious excitement of the evening.

'Isabella!' The relief in Christopher's voice is unmistakable.

Her soft smile warms him. She holds out her hands to him.

'Isabella, my dear. Thank God. I love you.'

Her smile deepens. Her eyes dance.

'You are a foolish boy still, Christopher, after all these years. You are a comforting person, all the same. A "nice" person,' she teased him. 'If I could not love you as you would wish, I think we have been very happy together. In spite of my many faults, I should like to think that I am forgiven.'

'As if there were anything to forgive, Isabella. With all your faults, as you call them, I love you still, with all my heart.'

'Dear Christopher.'

Hand in hand they go along the whitewashed corridor to the position in the wings from where Isabella will make her entrance for Act Two. The cats' concert of the tuning orchestra is discordantly perceptible at this point. Lisa, waiting, puts her arms about Isabella and embraces her. No quantity of make-up can hide the fact that she has spent the intermission weeping.

'Maman, oh maman.'

'My child, my child,' Isabella murmurs. Her face is soft as she holds the girl in her arms. 'There is no reason . . . .'

For some seconds they stand in this manner, until something of Isabella's calmness and fortitude enter into Lisa. The girl ceases to tremble; she breathes regularly; at last she smiles, a frightened little smile that goes in and out.

'So, once again,' says Isabella, 'think of no-one. Think of nothing except the music. Take a deep breath now, and go and sing.'

But even as she comforts the girl Isabella knows how easy it is to give advice, and how difficult that advice is to follow. She feels within herself the tightening of the diaphragm and the faint onset of apprehension she has never been able to conquer, try how she will, and with it the minatory needle-sharp pain under her breast which persists in spite of the all-too-short rest, and the tablet she has taken.

She grips Christopher's hand tightly, and puts her arm about Lisa's shoulder. She stands proudly, waiting.

The introduction to Act Two has a quieter beginning, although it works up into something of a fury, *vivace e veloce*. The first notes steal through the curtain, behind which the Ladies and Gentlemen of the chorus, almost choking from some minutes of enforced silence, lift their heads as old warhorses are said to do at the sound of battle. On cue the Stage Manager signals his underlings. On cue they grip their ropes firmly; they heave, and the curtains part with a velvet swish.

The Stage Manager gives a sigh of relief, and mops his brow with a very large handkerchief.

The music that wells up from the pit is still light in conception, evocative of the high spirits and light-hearted banter of the market place which now occupies the stage. Disguised as villagers, the members of the chorus adapt their movements and their singing to the increasingly intense rhythm of the music. They exchange snatches of musical conversation. They move about the stage with no apparent purpose, but in reality actuated by those unwritten laws which frown upon histrionic inactivity. To this end some artistic mind has visualised the grouping of largely primary colours so that the eye is not offended by the obvious lack of choreography. The scene represents some fiesta of importance; some village ceremonial, some rural interpretation of a national festival. It does not matter. The stage is brightly lit and well disposed. Wine is being served *al fresco* before the door of the *Cat-in-Boots* Inn; water is drawn from the well — that good and reliable stage property which, by its mere frequent use, has repaid its original price many times over. Gossips exchange the local news to the point of nausea; the young men boast and swagger; the young ladies flirt and giggle. The stage is alive with an interchanging kaleidoscope of movement and colour.

But it is a scene without inspiration. The music is pretty enough, the setting is all it ought to be, the choral singing is mechanically good. But there is no focal point. The introduction to the second act is an example of one of those well-known *longeurs* which occur in the best-conceived operas, when the muse nods, and technique is a little inadequate. They have been a stumbling block to many composers of otherwise fine music. Here is the weak point of Andelli's opera.

The audience listens, simulating attention, because there is little else to do. They are waiting for the re-appearance of Madame Weber. Since there has been no announcement of cessation, they presume that Madame Weber will continue to sing, and when she sings it will be effortless, perfect, and unbelievably beautiful.

Those who are privileged to be present this evening, and they are after all the people who matter, are therefore content. They consider themselves amply repaid, since they know beyond doubt that they are hearing singing which has never been, and can never be, surpassed. They know nothing of the unending heartbreaking practice, the private agonies of doubt, the denial of life and living which have gone to produce this perfection; they comprehend nothing of the inertia, the greed, the jealousy, the criticism which constitute the basic intolerances of those engaged in the profession of music. And if they had known, they would have dismissed them as artistic exaggeration.

What matters in the end is the result. The good citizens of Strelsau rejoice that a child of their city (so rumour has it) has reached such a pinnacle of fame as that now occupied by Madame Weber. Henceforth Strelsau, that city of cultural pretension, will by virtue of one woman have its place in the history of music. So wrapt are the good citizens in their contemplation of this consummation that they have temporarily forgotten the occasion for which such a *non pareil* was brought to them. The cause of it all, it might be said, sits above them framed by the glowing crimson of the many-looped and pendant curtains, his apparent beneficence bestowed on them all. They feel a little ashamed of themselves. It is not every day that one's Prince is crowned; one should not forget, even for a moment, so easily. A guilty glance in the direction of the Royal Box assuages many an uneasy conscience a little, although such a glance reveals nothing more than grim majesty on the one hand and raddled bovine beauty on the other.

'His Majesty appears to be unhappy,' they whisper behind their fans.

Certainly the atmosphere in the Royal Box is far from presenting that the spirit of joyousness which might be supposed to be almost mandatory on such a day. It is an atmosphere of hurt dignity, anger, disappointment, and bewilderment, sitting like a cloud over the head of His Majesty. It is, needless to say, highly infectious. It effects the phlegmatic lethargy of Her Majesty, and the demeanour of the Ladies and Gentlemen-in-Waiting, whose delicate airs and mincing gestures are quite dampened by its pressure. It is clear that as the evening progresses His Majesty has become a person of little account. Grand Patron of the Opera, Amateur of Music, nourisher of such poor plastic Arts as the city can produce, he has declined without protest into the symbol of Kingship, a threefold symbol of Politics, Commercialism, and Morality; a symbol the only function of which is to be, to maintain the steadfastness of his country in those three important areas, and to provide — Nature being co-operative — a future Grand Duke, thereafter to pass into Eternity, his passing to leave behind a line, or possibly a paragraph, in the history books.

The hour of his triumph has become the triumph of Madame Isabella Weber, a mere opera singer who has contumaciously rejected the condescension of a personal visit by the ruling prince. His face grows darker. From whence do these vagabond singing birds draw their pride, that it is as great as the pride of princes? Like a bear in a cage he stares about the great auditorium, seeking some explanation from the wondering smiling curious glances bent upon him for the barrier between him and them.

He tries to persuade himself that this is fantasy, the product of a morbid imagination fuelled by a febrile conscience. The more he contemplates this idea the more he fails to convince himself. He reminds himself that her place at the Coronation — reserved at his insistence — remained unoccupied throughout the ceremony, a situation studiously uncommented upon by officialdom. She has not acknowledged, by word or deed, his gift of magnificent flowers; she has shown no sign of having received his message; finally she has denied him entrance, as her King and Patron, to her room, an entrance which, were he of coarser flesh, he could have compelled by right and tradition. Grossest impertinence of all, she has stolen away the hearts of his subjects.

Hers is the fault, and hers is the loss. The honour he was to confer upon her, and honour that is scarcely less than her due, he will defer until such time as modest humility shall have informed untimely arrogance.

*

Pacing before the great fireplace during the unprofitable time between the end of the banquet and the departure for the opera, he had thought only to please by drawing gently together the fragile threads of friendship. All roads, he had surmised, however initially diverse their directions, lead eventually to a common point.

Today, tonight, was such a point, no less in time than in space.

Isabella. Our birthday. Our anniversary.

He indited and destroyed many messages before he finally reduced the longing to express himself to those few artless words. The heavy scratching of the nib jarred the oppressive silence of the small chamber into which he had withdrawn, disturbing the ghosts of memory locked by the gloom in corners, so that they peered out, grinned mirthlessly, and clapped their silent bony hands. They gesticulated in the ebb and flow of the shadows.

How would she receive such a message, he pondered, stamping it with the purple seal as if it were a royal *directif.*

How will she receive such a message? The ghosts mocked him and grinned again.

How would she reply?

How will she reply? They twist themselves into contortions of malicious ecstasy.

She will not reply, they mouth at him soundlessly. *She will not reply, not reply . . . reply . . . .*

She must reply. Courtesy demands it.

But there is no reply.

The ghosts withdraw into their Stygian limbo. They are satisfied. They grin, they grimace, they clap their fleshless hands with a dry rattling that rouses His Majesty from the depths of his bitter reverie to a startled awareness. A rattle on the timpani heralds a reorganisation of movement on stage. The attention of the audience is concentrated on that slight black figure on the podium, and the peculiar hop, skip and jump that is characteristic of all Andelli's movements in moments of tension. His rounded back, forward-thrusting head and flying hair have been caricatured, sympathetically or otherwise, in music journals all over the civilised world. He rarely recognises himself.

His contortions are an animated illustration of inherent exhibitionism struggling for pre-eminence with mastery of craft.

His Majesty considers him dispassionately as his antics increase, persuading his musicians to the requisite tempo, preparation and resolution that heralds the imminent appearance of Madame Weber in the second act.

His Majesty mentally prepares himself, but with little hope of resolution, for the battle of wills that will surely begin again with Madame's entrance.

Beside him Her Majesty selects another bonbon, and yawns while it is still halfway to her mouth. The Ladies-in-Waiting nudge each other, and exchange sly grins. If they have no interest in persuading her to eat bonbons, neither have they an interest in persuading her not to eat them. It would seem that they even derive some peculiar pleasure from guessing how many bonbons Her Majesty can or will eat in any given length of time. They even enjoy her periodical attacks of nausea and dyspepsia, but they do not enjoy the fits of bad temper which precede them.

His Queen. His wife . . . Amelia Viktoria grinds the bonbon between her discoloured teeth. *His* thoughts are reflected upon her bovine features, and she glances sideways at her husband.

It is understood that Madame Weber adopted her, Your Majesty.

As a baby . . . .

His Majesty is again agitated by a succession of memories, a suspense of doubts. His thought revolved about the physical appearance of the artiste Lisa Weber, and the resemblance to Madame Weber. Adopted . . .?

Isabella, *our birthday . . . our Anniversary . . . .*

The ghosts grin gleefully and gesticulate in the corners of the mind.

Our Anniversary.

She calls herself Lisa Weber . . . .

An element of certainty begins to drive out conjecture. His Majesty's hands tremble; perspiration starts upon his brow, a perspiration which has nothing to do with the heat within the theatre, but has its origin in a sudden warmth in his frozen heart. He feels that he may be on the point of solving the enigma.

Our Anniversary.

On such a night as this, those many years ago, with the winter wind howling about the square and hurling the snow into corners and blanketing the bronze nakedness of the Grand Duke Stanislaus's horse, he experienced the onset of an emotion which brooked no denial. At the age of twenty-one, aristocratic, arrogant, accustomed in the wild years of his nonage to take what he wanted at all times, he waited with the impatience of youth and an increasing irritation for an end to the opera which had been put on with all ceremony in honour of his coming of age.

Before the applause, prolonged and verging on the hysterical, had shown any signs of subsiding, he had made his way, brushing aside gracelessly all sycophancy, to the dressing room of Fräulein Isabella Weber, the rising new star in the firmament of music.

Without announcing his presence, he threw open the door, to be confronted immediately by the disapprobation of the formidable Maria. His impetuosity lost a little of its force.

With a sour note in her voice, Maria announced: 'His Highness the Grand Duke . . . .'

Isabella raised herself a little from her chaise-longue.

For some time there was a constrained silence. Then at last he found his tongue.

'How beautiful you are, Fräulein,' he said.

Isabella stiffened.

'Sire?'

But it was a reprimand rather than a question.

'How beautiful you are,' he repeated, in no way abashed, 'and how wonderfully you sing.'

Isabella rose to her feet, slowly, gracefully. She was as tall as he, he realised, when seen under normal circumstances. Her hair was red, and her eyes green.

She considered him for a moment. It was not a difficult thing to do, for he was quite handsome. She softened. She smiled. She made a slight curtsey.

'I thank you, Sire. I sing the best I can. Wonderfully or not, it is scarcely for me to say.'

'You do yourself an injustice, Fräulein Weber.'

Isabella lowered her eyes.

'Your Highness is too kind,' she murmured.

From within the wardrobe Maria sniffed loudly.

'One must speak what is in one's heart, Fräulein. Never have I heard such a voice, and I dare swear I shall never hear such another. Such singing, and such beauty with it, deserves recognition. I personally shall attend to it.'

Isabella waited.

'Unfortunately, I am awaited at the Castle. There is to be a Ball.'

'Somewhat lacking Cinderella,' Maria muttered, her voice discreetly muffled.

'You must not delay, your Highness,' said Isabella. 'It is already late.

'I shall never forget my anniversary, Fräulein Weber.'

Isabella smiled wryly.

'For your Highness, a grand holiday, I have no doubt. For me, a working day. It is by chance my Anniversary also, but alas, no brilliant concourse of friends awaits me. My — duties, shall I say, are finished.'

His Highness laughed joyously.

'In that case, Fräulein, we have a double duty to observe. My excess of ceremony; your lack of it. And observe it we shall. Such beauty and such talent,' he repeated himself, 'cannot go unappreciated. Tomorrow, Fräulein Weber, with your permission, I shall call upon you, and you and I shall celebrate more intimately our double anniversary.'

He bowed, very gravely. Isabella curtseyed, very low.

'Your Highness,' she murmured.

'Until tomorrow, Fräulein. *Mit Gott* . . . .'

Maria, emerging from her wardrobe closed the door behind him with a scarcely perceptible bang.

She sniffed loudly.

'*How beautiful we are*,' she mimicked. '*How wonderfully we sing*. So, another fly has thrust itself into the web.'

The colour rose in Isabella's pale cheeks.

'He has to say those meaningless things, Maria. No one would take them seriously.'

'*With your permission, Fräulein*,' Maria sniffed again even more pointedly. '*We will call upon you*.'

Now Isabella blushed furiously.

'You embarrass me, Maria. He is the Grand Duke. His wish is our command.'

'Since when has anyone commanded Isabella Weber? You are a fool, Madame, begging your pardon. But he is such a grand fly, and no doubt an honourable young man. After five years of important work and study you have to fall in with a brash Ruritanian princeling. It is too much.'

Isabella's eyes flashed ominously.

'Now I begin to lose patience. You talk too much, Maria.'

'Yes, Madame.'

Isabella laughed bitterly.

'My life is my own, I think. I shall order it as I wish.'

'Yes, Madame. As you say, your life is your own. Now we shall dress, if you please. The day is over. *Träume süss* . . .'

Late in the afternoon of the following day the cold wind had sobbed itself to rest, and the snow had ceased to fall, having accomplished its task of covering the winter land with a pristine warmth-preserving blanket. Forgotten was the solemnity of a royal anniversary, the revels of a Ball where even the swirl and swish of bejewelled women and the gyrations of poker-faced men can quickly pall. Gone were the garish lights of the theatre and the sluggish moving tide of perspiring faces, the hoarse voices croaking their uninformed approval of matters beyond their comprehension. Gone was that interminable scraping of cat-gut and blowing through brazen pipes which is the nearest that man can get to the music of the spheres. His Highness drove his sleigh with a carefree hand and heart through undefined woods and forests, and as he drove he talked of this and that and other nonsense, for the sheer pleasure of hearing a gentle drowsy reply. The caress in Isabella's voice, so distinctly different from the exquisite singing tone, bound his heart with hands of iron.

The sky grew darker. The first stars appeared. Isabella watched his face as he drove, so young, yet so old. So traditionally arrogant, and yet so thoughtful and protective. She stirred a little, and he tucked the bearskin wrap more closely about her, leaning over her. He boasted of the bear he had shot when he had been a mere stripling, how proud he had been, and how sick afterwards. But it had truly been a monster of a bear, savage with hunger. Isabella snuggled deeply into the pliant folds. The keen night air whipped her cheeks into a glow; she was lulled into drowsiness by the jingle of harness and the muffled pounding of hooves in the snow. She dozed through the darkness of the pine woods, through the cold stillness of the white hills, under the dome of the bright stars, until at last they came to the inn, its door already open in welcome, its red curtains glowing with an inner warmth.

'Have no fear, Isabella,' His Highness reassured her. 'The house is mine. Here we are at home. The host — my former *major-domo* — is paid well enough to be the personification of discretion.'

With some trepidation Isabella stepped over the threshold into the warm room where, horror upon horror, a fiddler, apparently satisfied with his excruciating tuning noises, began immediately to play, at first *piano*, astringently, as though to erase the mind of recent too sweet thoughts.

The orchestra takes up the theme; the ladies and gentlemen of the Chorus regroup themselves into yet another amorphous pattern, leaving the centre of the stage clear. They intone what has become known in operatic circles as Andelli's *Festgesang*. It is based on a folk theme which reaches back into the dim ante-chamber of history when gods were indeed gods, and walked the Earth. Andelli has polished it, given it a local habitation and a name. It is simple and beautiful, and everyone sings it or vocalises it, according to individuality, everywhere.

Andelli's beat at this point is firm and precise. He is leading his musicians with some deliberation to the point of Madame Weber's entrance.

The audience is still, tingling with anticipation. '*Now*,' they say to themselves. '*Now comes our Isabella again, in four bars, in three bars, in two bars* . . . .'

Lisa presses Isabella's hand. 'Please take care, Maman,' she whispers.

Isabella returns the pressure. 'Courage, my child. What is to be, will be.'

Maria removes Isabella's wrap. Isabella takes a step forward. Andelli glances upstage, his baton poised.

Now . . . .

With such smoothness and faultless execution that no one realises, for a moment, that she is now singing, Isabella enters into what has been called, for want of a better description, the *Obbligato for Voice, Chorus, and Orchestra.* It is not, of course, an *obbligato* in the technical sense of the word, and it is sufficient to say that this particular passage called down upon Andelli's head after the first performance a great storm of criticism on account of its uncompromising modernity, its neo-classicism, its unashamed romanticism. No stick was too short or too long to beat the dog with. It had no place in an opera; it was completely out of character; it was symphonic rather than operatic. Who was this young upstart, this reactionary, this ignoramus, that he should set at nought the precepts of musical composition?

Needless to say, it is the perfect vehicle for the display of Madame Weber's unique artistry, and remains one of the world's great arias.

The audience is completely absorbed, elated even. They perceive no trace of the rumoured recent indisposition. 'Their Isabella' is singing as she has rarely sung. She is singing too well, the thought occurs to Christopher, tense in his corner. She is singing as if this performance were likely to be her swan song. He cannot shrug off the feeling that there is tragedy in the air. He is oppressed by a nameless fear. He is still convinced that he was right to oppose this return to Strelsau, that nothing good could come of it. Sleeping dogs are best left sleeping. The same opera, the same principals (with the exception of Lisa), the same conductor, the same theatre, of necessity; and, of course, His Highness the grand Duke, now His Majesty Karl Ernst Maria Michael Frederik, King of Ruritania, accompanied by Her Majesty Amelia Viktoria Konstanzia Camilla naturally — what folly, to assemble such a cast for a drama which is neither of the *seria* nor the *buffa*, but partakes of both. What exquisite folly!

Why had she behaved in this way? A desire to demonstrate to the good citizens of Strelsau that Isabella Weber was no longer a backyard hen scarcely held water. The passage of time alone had destroyed this myth completely. A desire to repeat a success in the place of its birth is not a logical solution. In twenty-one years the musical world has recovered from the shock of its innovatory ideas, and has passed on to other new things. Finally, most inexplicably, she has responded for the first time in her career to the demand for an encore.

Christopher regrets his inability to understand Isabella. He regrets that he has never been allowed to share her complete confidence. He regrets that her mercurial mind has always been well in advance of his more logical but pedestrian efforts at cerebration. Perhaps, he considers ruefully, the fault lies in himself. He has not been able to help her because he was not aware that she needed help. Even so, he has loved her sincerely, and served her faithfully after his fashion. He recalls the agony and the ecstasy of love given vainly; the times of happiness and delight; the darker times of doubt and despondency and the joy when they laughed together; the misery when he was forced to realise that he was to her no more, in terms of her beloved theatre, than a reliable 'prop'.

Standing discreetly in the wings, as he does at every performance, he sighs, and consults his watch with the nervous gesture that has become habitual with him. The second act is already advanced by some thirty minutes. All is, for the moment, going well, but he cannot account for the great feeling of depression which has fallen upon him. Perhaps it is in part the result of the Doctor's unqualified pessimism; perhaps it is, more basically, the result of the failure of those nearest to her to recognise the pressures of her chosen way of life on the mind of a mere child. She had not wanted for care and affection; she had lacked neither instruction nor guidance. But her unique personality had precluded understanding. So perhaps after all it was not his fault. They had all, his father, Herr Weingartner, even Matthild' and himself wished nothing but good.

*

The result of their efforts and good intentions lay on the chaise-longue, a still, pale child with no sign of life.

There was nothing one could do. She would live or die, according to the will of God. That was the Doctor's first opinion. He personally supervised her removal from the chaise-longue to Matthild''s most comfortable bed in the spacious room overlooking the Stanislausplatz. From here, should she open her eyes, she would see immediately the Gothic facade of the cathedral, the snow-covered rump of the Grand Duke's horse (that unlovely monument!), and an oblique view of the front of the Opera House.

There was not even a flutter of the eyelids. She lay deep in a coma in which she scarcely seemed to breathe. Each day the Doctor, waited upon and observed by Matthild' — an excellent nurse, he complimented her — examined the patient. He tested her pulse against the great turnip of a watch that he carried, checked her pulse and respiration, took the temperature of the unmoving limbs, lifted the eyelids. It was simply medical routine. The Doctor made no comment, except to instruct Matthild' to keep the patient warm and quiet. At the end of his visit, having taken a glass of Josef's good German wine, he would shrug his shoulders, bid Matthild' *zum Morgen*, and trot off, conducting some inner conversation with himself.

'What did they think they were about,' was the burden of his complaint, 'to submit a child to such a routine?'

Perhaps those two — by which he meant Herr Smith, a nincompoop, and Herr Josef Sigismundus Weingartner, a well-meaning bully, had by now realised just how much they were to blame.

He was not to know that they had indeed searched their hearts privately, but that they kept their frigid thoughts to themselves, for they could not begin to imagine how they were in error. The days went slowly by, their passing measured chiefly by the strokes announcing the hours reverberating about the snow-covered square, trembling into silence as they reached the threshold of the room where Isabella lay. For the first time in many years the daily routine of Herr Weingartner's precise but comfortable house revolved round a person other than Herr Weingartner. He was to a large extent ignored by both his sister and her maid Maria, the latter of whom, being some three or four years older than Isabella, had taken the child immediately to her capacious heart. The frailty of the unconscious child seemed to have released a fountain of maternal feeling which surprised Matthild' also. It was a feeling she had lost hope of ever experiencing. Hour by hour she sat by the bed, stroking the lifeless hands, smoothing the no-longer shining hair.

When he was not otherwise occupied, Josef sat at the other side of the bed. He rarely spoke to his sister, and even rarely looked at her. He simply stared continuously at the pale, immobile face, as if by the sheer effort of will he could restore it to life and movement. Sometimes his lips moved, as if in prayer. He was not a particularly religious person, since he had encountered few difficulties in life, and it is possible that the humility demanded by the need to formulate a prayer was more than he could achieve. He seemed to have aged somewhat suddenly in the last few days. He did not eat or drink with such gusto, and found little solace in his tobacco. Morning and afternoon he enfolded himself in layer upon layer of clothing and plunged into the snow to visit the homes of his lesser 'singing birds'; he had little heart for these journeys, but he was a conscientious man, and took his duties to those less gifted seriously. The house remained quiet, Isabella was not disturbed, and nature continued on its healing course.

Christopher felt himself to be of little use during these dark days. Neglected by his father, although not deliberately, he was sure, he could only pass the time aimlessly, his thoughts and feelings so confused that nothing he did seemed to have any purpose. It was plain that, although Josef had accepted him as an unavoidable addition to the group that was becoming by degrees as intimate as a family, he was by no means in favour with Matthild', who treated him quite simply as another cross to be borne. He was discouraged from visiting Isabella too frequently, and he had to rely on brief communication with Maria, who having something of a *tendresse* for this strange English boy, passed on whatever news she had concerning the progress, lamentably little, of the patient. She was woman enough to know that, although something had gone out of him, something else had entered in. At the moment when she opened the door to him, her own face bright and beaming, and had whispered the news to him that Isabella had at last moved of her own accord under the covers, she knew that he was no longer a boy.

After seven days Isabella had actually moved, she told him, and her breathing was deeper and more regular. The Doctor, sent for immediately by Matthild', was with her now, but alas, poor man, there seemed to be nothing he could do, except to go through the usual motions of his profession.

'It is too soon,' he said, when Christopher would have asked him a hundred questions. 'It is too soon for me to express an honest opinion. Nevertheless, the signs are encouraging. She will recover, ultimately.'

For Christopher that was enough. She would recover. His love would return to him. Once again that young urgent beauty would delight him and the world by simply *being*.

For Papa Smith and Josef Weingartner there were other darker thoughts. While Papa Smith was in the first place sincerely concerned about the ultimate recovery and future health of Isabella, it must be said that Josef's thoughts were more basic; would her voice be permanently affected? Would she indeed ever be able to sing again? Would she ever again be strong enough and healthy enough to achieve and sustain the career he had in his heart mapped out for her?

They both had to wait for an answer.

For some time Isabella did not move again. It was in fact twelve days after the initial collapse before the first positive signs of returning consciousness appeared.

*

Standing in the wings, with the draught playing about his feet and ankles, his eyes and ears alert for signs that all is not well, Christopher recalls those unhappy hours of waiting and watching, those hours of despair when even the brightest day has a darkness of its own. Surely it was the veritable will of God that she should come out of the shadow of that terrible Valley to delight the world with her singing. Her 'family', as she was to call them, watched and prayed, each in his or her own way. Their prayers were undoubtedly answered, as is clear to all on this wonderful evening. But still the shadow persists, enfolding Christopher in a black cloak of apprehension.

Even before Isabella begins to sing, Christopher flinches at the thought of what effect upon her this almost literal recital of the tragedy of her life will have. Again he is astounded by the remarkable prescience shown by Andelli in the composition of this opera. Try as he may, he cannot understand how the life of a person can be foreseen with such accuracy — allowing for the discrepancies arising from poetic and musical licence. What understanding, what intuition had Andelli that he could put into music all the life of a girl whose life had scarcely begun?

'I am Italian, and I am a musician,' he bombasted. 'I see into the heart, my friend, I understand the soul . . . .'

'Nonsense!' Christopher could not prevent the exclamation.

'But it is true, *amico*,' Andelli protested. 'I watch, I listen; I think. *Ecco*, I have it. There is the character, and there, as a consequence, is the life. *Che sara, sara*. As you might say: *what is written, is written*. Isabella has an intriguing personality. She inspires me, and the music writes itself. Phrase upon phrase it unfolds like a flower in the sunshine. The grand theme takes shape in my mind. But alas, my friend, do what I will, it comes in the costume of tragedy, as indeed does the best of music, I think. The tragedy in my music must reflect the tragedy in the life of my heroine. Why do I imagine the life of Isabella in my tragedy? I cannot tell; she seems to be happy enough at the moment. I watch, I listen, and I feel tragedy. That is all. Isabella shall be my heroine, and she shall sing my music. But I shall not think of her as Isabella. That would not be at all correct. You do not mind, Fräulein?'

He rose to his feet as she entered the room, and sketched a bow.

'What do I not mind, Signor Andelli?'

'You are to be the heroine of my opera, Fräulein.'

She curtseyed in her turn, deeply, satirically *à la mode*.

'But, I was explaining to Christopher that I shall not call you Isabella. That pretty name would not be at all compatible with my conception of the opera.'

'What then shall you call me?'

'I shall call you — her — Lucia. No, Margharita, Elvira . . . pouf, what stupid names women have. Ghismonda, Gretchen . . . stupid, stupid.'

He snapped his fingers as if he were a prestidigitateur calling up names from the ambient air.

'Judith, Juliet, no . . . Isabeau, no, indeed not. Louisa, Marta . . . I am well acquainted with a singing animal who goes by the name of Marta.'

He swept back his long hair in Italianate desperation.

'*Cielo*. What can she be called?'

Isabella and Christopher waited, in mock fear.

'I see her as high born, calm and self possessed . . . and . . . .'

'Ox-like,' suggested Christopher, his face straight.

Andelli fixed him with a glittering eye.

'You are pleased to mock me, Signor Christopher. I have it. She shall be *Cordelia*. Has not your Shakespeare such a heroine?'

Christopher grinned.

'*Her voice was ever soft, gentle and low, an excellent thing in woman*,' he quoted.

Now Isabella turned upon him in fury.

'You beast, Christopher.'

He caught her flailing fists beating impotently against his chest.

'I was complimenting you, Isabella.'

She swept him with stormy eyes.

'You were not. You were mocking me. I do not screech or scream. I think Cordelia is a lovely name. Apologise, Christopher, this instant. Signor Andelli, make him apologise.'

Andelli shrugged his shoulders and spread his hands.

'Would that I could, Signorina. But discretion is the better part of valour, and alas, Signor Cristoforo is bigger than I.'

His suavity was as impregnable as a rock.

'I hate you both,' stormed Isabella.

'That is impossible, Signorina,' Andelli declared. 'That is manifestly impossible. We are your best friends. We love you. How then can you hate us?'

'That is nonsense,' put in Christopher. 'You do not love her. You Italians are all the same. You have only to see a girl, and it is love for breakfast, lunch and dinner. You think love, you dream love, and you have no idea what love is. You talk about it too much. *Ti voglio bene. Mi voli bene*?'

He mimicked an Italian talking of love. Isabella giggled, her usual reaction when Christopher was at his most English.

'Si signor,' Andelli agreed gravely. 'Such is our fault. We talk too much about a most important matter. Perhaps our feelings are as deep as those of you English, who say nothing at all. Perhaps you too do not know what love is.'

'I am sure that is true,' said Isabella.

Andelli bowed again, and quoted: '*O Pietosa colei che mi soccorse*. Oh blessed she that stooped to take my part. *E tu cortese che ubbidisti tosto, Alle vere parole che ti porse*!'

Isabella curtseyed in her turn.

'I thank you, kind sir. But I will not be your Beatrice. I am Cordelia. I am high born; I am gentle, I am even submissive. Do you remember?'

'Cordelia. I am Cordelia,' murmured Andelli. He hummed a phrase tentatively. 'Yes, that is how it should go. Sing, Cordelia . . . . *Mi chiama Cordelia*.'

Isabella sang. Her notes were a true repetition of his, but cleaner, more pure, bell-like. Twice she sang the phrase, at first softly, as if in a dream; then on a crescendo, with increasing strength.

'*Mi chiama Cordelia*.'

Andelli hummed the phrase again, and added a further phrase, which Isabella repeated.

'*Mi chiama Cordelia . . . . Altro nome non esiste . . . .*'

'Beautiful, beautiful,' exulted Andelli. 'But softly, hopelessly. There can be no expression at this point.'

*

On his precise downward beat Cordelia enters upon that beautiful aria which recounts the humiliations and sufferings of a noble soul. The notes glide needle-sharp over the orchestra.

Son orfana, she sings, and every note is clean and pointed. Effect follows cause, inevitably and rationally, building up to a glorious climax of sound and feeling that is almost more than the heart can support.

'That is how I hear it,' Andelli had declared, and that is how it is. Not for the first time Christopher is compelled to admit that Andelli is in all ways a master musician.

The House is silent; it is as though the sadness of human conceit were being stripped away layer by layer to expose the ultimate agony of the naked soul. The torment of Cordelia's face is terrifying; her body is twisted into the suffering lines of a woman who has borne the heaviest punishment that life can inflict.

Note by note the theme builds up to its climax. Andelli's eyes and hands plead, persuade, command. The response is magnificent. His lips smile briefly, as *piano, crescendo,* mount to the terrible *forte* of the agony which twists all hearts. At this point Cordelia's submissive sorrow turns to anger as she relates the trickery by which she has been betrayed. She stands erect and defiant, facing her audience. There are mutterings, and uneasy shiftings in the velvet seats. Conscience is an uneasy companion under such conditions.

On the rostrum Andelli is completely possessed by his creation. The orchestra plays as it has never played before. Christopher shudders at the awesome perfection of Isabella's singing. There is no stage, no trumpery canvas scenery. There is simply a woman, Everywoman, All-woman, stating her sufferings before the Judgment Seat, to the accompaniment of music made from the silver strings of Hope, the tarnished golden strings of happiness, and the dark strings of suffering, woven into a pattern of sound such as might emanate from the spheres.

It is almost the climax of the aria. Apprehension still sits heavily upon Christopher's shoulders. He waits, and wonders whether the next two hours will pass without untoward incident. In two short hours whatever purpose she had in mind will presumably have been fulfilled, and it will all be over.

Still she sings, but the tone is quieter, sobbing, almost. A dying fall. The percussionist prepares his stick, raises it . . . leans toward his metal tubes. Strikes . . . .

The painful sound vibrates about the square, indicating that at least some minutes of the two hours have passed without incident, and that the smooth passage of the remaining minutes might reasonably be hoped for. Christopher ignores it. Long ago he lived with it; it was daily the background of his life and his love. It has little meaning now. There is only the voice of Isabella, and a gathering sense of loneliness.

As she sings the last note of the aria Cordelia sinks to her knees. For some minutes, during which one might have said a *Hail Mary*, there is a palpable silence throughout the House. A faint movement of the hands, a whisper in a far corner, and the tension is released. The roar of voices rises to the ceiling; hands are clapped, feet are stamped, flowers are thrown. Such a scene has not been witnessed before. Only in the Royal Box is there no reaction, unless one accepts the stony face of His Majesty, and a yawn from his consort Amelia Viktoria, as a sufficient reaction.

The uproar continues. The ladies and gentlemen of the Chorus stand frozen in their places, while Cordelia crouches, an unmoving bundle, in the centre of the stage.

'God has heard her sing.'

The voice is still sardonic, and is supported by a sardonic rictus in place of the ready smile of former years. The words have the same effect on Christopher now as they had then.

'Your words are still ill chosen, Herr Doktor . . . .'

The Doctor shows his teeth. In the diminished reflection of his peculiarly-shaped lenses Christopher can see a miniature Isabella in duplicate, framed in the narrow space between two flats.

'God has indeed heard her sing,' the Doctor repeats. 'So has His Majesty up there,' he continued, looking upwards. 'For His Majesty this performance is a twist of the knife in an old wound. For God it is a problem in human frailty. You would have thought that He would have arranged matters better.'

'You talk in riddles,' Christopher informs him.

Again the Doctor's ancient face cracks into a thousand creases.

'There are no eyes so blind as those that will not see.'

Christopher stares at him. The roar in the House continues, greater than before.

'Isabella,' they call ceaselessly. 'Isabella . . . bella . . . bella . . . .'

'Why are you not in your seat?' Christopher demands to know.

'You, of all people should know,' the Doctor replies. 'Am I not the official — what is your word? *Sawbones*? — of the Theatre?' He is manifestly pleased with his familiarity with the English idiom. 'Someone must care. You, my friend, would do well to pay a little more attention to the welfare of the lady. There are regrettably all too few such.'

Unaccountably chilled by this remark, Christopher turns his attention to the stage. The applause continues, hysterical, melodramatic. So long as it continues, Cordelia remains silent, unmoving, her whole posture one of dejection, utter exhaustion. If it were not simply a most finished theatrical performance, it would perhaps be unbearable. As the music falls to its resolving chord and that consequent moment of silence which is now recognised by writers on musical affairs as Andelli's signature, Cordelia closes her eyes.

Somewhat shamefacedly, the audience falls silent. Too late it is realised that one cannot applaud such perfection, any more than one can stand before the *Mona Lisa* and clap.

Andelli stirs into action. The music soft and slow, a mere accompaniment, swells into a marked bass. The inn-keeper, attended by his son, who carries a small barrel upon his shoulder, approaches Cordelia. He stops, bends, takes her by the arm.

'In the name of God, child,' he sings. 'What do you do here? Whence do you come?'

There is no response from Cordelia.

'What is your name, child? Who are you?'

Now Cordelia looks up into his broad peasant face, and is strangely comforted.

'They call me Isabella,' she replies. 'Other names have I none.'

In his corner Christopher gasps, and sees Isabella's eyes open wide in panic. No one moves in the shadowy pressing circle about her. No sign indicates that an unprofessional slip has been made. The music continues. Seeking Andelli's face in the darkness below, Isabella feels rather than sees his glittering eyes upon her, willing her with all his strength to regain her self-control. About the audience there is the silence of the grave. Not even the rustle of a programme disturbs the tension of the scene. The plaintive harmonies throb about them, and about Isabella as she struggles to the surface of reality.

In a voice so small that it can scarcely be heard, she repeats her line:

'*Mi chiama Cordelia. Altro nome non esiste . . . .*'

'*Cordelia. That is a pretty name, and whence do you come, Cordelia*?'

She peers this way and that through the mist before her eyes, endeavouring by some transfer of thought to convey to him that under the heavy snow-laden sky, beyond the lowering roofs and the dark alleys foul with the reek of primitive humanity, beyond the cluttered courtyards and the broken casements, lies what she once thought of as home.

He cannot understand, but he is moved with compassion. At his signal, the younger man gathers Cordelia into his strong arms, and together they go into the inn, the assembled peasants making way for them.

Piano, piano, the following theme begins to unwind itself. A perspiration of relief gathers on Andelli's forehead.

Crescendo . . . . A gentle *crescendo*, and the phrase curls itself about the heart.

'Her name is Cordelia. She has no other name . . . .' The peasants chant in surprised unison, asking themselves how this can be.

The audience moves restlessly.

'*Was ist los*?' several of them ask, looking at their neighbours in wild surmise, but receiving no answer.

Others think:

'*What an actress. What a consummate actress.*'

But there are tears in Christopher's eyes.

*

When Isabella opened her eyes, and kept them open, on the twelfth day, and gazed wonderingly at each in turn of the four about her bed, they themselves were so worn out with the strain of waiting that at first they could not realise what had happened. They were so incredulous that they could do no more than remain unmoving, not daring to make a sound that might undo what good was being done. Isabella lay quite still in the huge bed. So huge was the bed, and so small was Isabella that she was all but lost in it. She lay like a doll, her face shrunken and wasted. The long gold-bronze hair framed a face in which the cheeks were so hollow and the bones so high that the nose stood out like the beak of an unfamiliar bird. Isabella was no longer beautiful. But she was Isabella, *his* Isabella.

It was late afternoon, Christopher remembers. The room — hot and oppressive because of the necessity of keeping the stove continually fed with logs — was already beginning to grow dark. Outside the snow lay deep on the Square, on the steps of the Opera House, on the rump of the Grand Duke Stanislaus's horse; and in the garden, the neat flowers beds and the subtle perfumes — the result of Matthild''s unrelenting attention — which had so delighted Isabella, were completely hidden. During the seemingly interminable days of watching and waiting, when the winter sun was shining, the snow was bright and sparkling. In the twilight it now reflected only a pallid watery light on the ceiling and so over the bed and the four faces intent upon it.

They leaned as closely as was compatible with good manners, searching in the green luminous eyes for some sign of life, some movement of recognition. For some fearful moments there was no movement, the eyes were simply open, staring at the pale, watery-looking ceiling.

In the stove a log fell suddenly, startling them all. Isabella moved under her covers. She moved her head from one to the other of their faces until she came to Christopher. She essayed a smile.

'Why were you dreaming, boy?'

The miracle they had prayed for had happened. From that pitifully thin throat had come the most beautiful sound they had ever heard.

Christopher blew his nose violently, and under cover of that operation he dried the tears in his eyes and on his cheeks. The rather vulgar noise he made broke the spell. Matthild' cast a reproachful glance at him, but the reproof, if such it was, was tempered with a dry kindly smile. There were times when Matthild' was almost human. Josef and Papa Smith stirred, clumsily, jerkily, like puppets suddenly called upon to perform. Their eyes wrinkled as they exchanged timid smiles. Tentatively they each stretched out an arm; hesitatingly, then firmly, they grasped each other's hand. Warmth began to seep back into their cold hearts.

Isabella will recover, they thought in unison, and their thoughts expanded about this idea as a flower opens in the sun. She would become well and strong again. She will sing again to charm two old men in the Autumn of their lives. They both heard in their minds that divine voice as it soared and floated to Heaven, so beautiful that it could make the angels jealous.

Together they turned towards her bed.

Together, as though it had been planned, they said: 'Isabella.'

Isabella had closed her eyes, and was sleeping naturally and peacefully.

When the Doctor appeared later in the evening, a medical and social visit combined, he received their eager news with a gravity which somewhat dampened their new-found happiness. Again Isabella was vaguely awake, and his examination was longer than usual, more detailed and precise, and watched over by Matthild' with pursed lips.

In the salon Josef and Papa Smith found little in his subsequent opinion about which they could immediately rejoice. They had no doubt, in their innocence, half-expected some medical wizardry to have restored Isabella to normality again. Nothing of this possibility appeared in the doctor's report to them.

They sat meekly while he spoke to them. He pointed each period with a sharp glance at each of them over his half-lenses, and a tap of his stethoscope on the palm of his hand. He requested that they attend to his words, for it was extremely important that they should understand everything.

She would probably make a full recovery — as far as the word was understood by those who had not had the advantage of a training in medical science — but it did not follow that she would be able to remember details of her life before her illness. There would probably be a partial disorganisation of memory for the rest of her life, which might also be accompanied by occasional — temporary — loss of speech.

Acting on the principal that for sensible people it is best to anticipate the worst, he spared them nothing. He was unmoved by the sad change in their expectant faces, and by the sob from Matthild' as she turned from him, her apron to her face. In such cases as this, he continued, one could not reasonably specify the extent of the nervous injury.

He paused. They wagged their heads in sorrow.

The Doctor pursed his lips judicially. He regarded them with his head on one side like an uneasy sparrow. He tapped his palm with his stethoscope again, as though, having cast them down, he was debating the advisability of lifting them up again. Although he considered that they bore joint responsibility for the present situation, he knew that they could never change their natures. He knew that the greatest unhappiness can be the result of the best of intentions.

'On the other hand,' he continued, a little more briskly, after a prolonged pause, 'she may make the completest of recoveries. Physically she seems to be in a hopefully reasonable condition, although extremely weak, naturally. The heart is quiet . . . .'

'What a wonderful phrase that is,' thought Christopher. 'The heart is quiet . . . .*Ruh' ist der Herz* . . . .'

'. . . and functioning normally.'

They waited in silence for his next words.

'Whether she will sing again, which is what you are waiting for me to tell you,' and he threw a malicious little glance at them, 'whether she will ever have the strength to sing again as she and you would wish, I cannot tell you. It is something which depends to a great extent on you.'

They babbled together.

'What can we do? We will do anything to help.'

He could not say: Search your hearts for a little of the love you profess. It was not for him to judge them, and he was quite sorry at their woebegone expression.

'It is rather a question of what you cannot do,' he told them.

They stared at him in unbelief.

'Isabella must be allowed to return to life in her own time and at her own pace. Any untoward stimulation would be most disadvantageous, and, in fact, I would accept no responsibility if she were to be disturbed in any way. Rest and quiet are still the essentials. For your part, an atmosphere of caring, love, and understanding is all that is necessary for some time to come.'

They continued to stare. They could not understand how such a negative attitude could be of any help. They wished to pour life-restoring liquids coloured red, green, and blue into her by the spoonful every four hours. They wanted to cosset her and pet her and continually wrap her around in warm blankets with hot bricks at her feet. They wanted to describe to her how ill she had been, and how, thanks to their vigilance, she was getting better. They wanted to tell her how much they loved her, and how everything was going to be much more interesting and amusing. They wanted . . . .

They were sharply pulled up in the middle of their catalogue of good thoughts. They gasped at the Doctor's impertinence. They protested.

'You are no better than children yourselves,' he said.

Their faces became solemn again.

They agreed that perhaps they had been too excited. But if he could only understand how they had suffered to see the little one lying there so still, so pale.

'Under such attentions as you suggest,' he almost snapped at them, 'she is like to lie there much longer, and become paler still.'

They nodded their head dolefully.

'As for your coloured water . . . . Medicines are the resort of the ignorant. They do no great harm; neither do they do much good. At present the body cannot deal with them, and the stomach will reject them.'

Without being aware of it they again wagged their heads like old councillors in committees.

The Doctor repeated his strictures.

'She is not to be moved.'

They nodded, to show that they understood perfectly.

'She is not to be stimulated in any way.'

They shook their heads.

'You must not attempt to speak to her unless she has been fully awake for some time.'

Their mouths opened in protestation. But they recollected themselves in time. They shook their heads very decisively.

'You must reserve your love and sympathy for later, when it may, or may not, be appreciated.'

'She has all our love,' Papa Smith told him. 'There is no one we love more in the world.'

'That I can believe,' the Doctor was persuaded. 'Unfortunately many far from admirable things are done in the name of love.'

While the two men were seeking to understand this remark, the Doctor turned to Matthild'.

'When she is able to take it, a little plain food will do no harm, beginning with the liquid variety. A nourishing soup, for example. You should be able to arrange that, Fräulein. It is not a very complicated business.'

Matthild' bridled. She sensed a hidden sarcasm. The provision of good food was at once her *métier* and her pride.

Look at her brother.

The Doctor looked, and shook his head.

'My friend Josef is no doubt well cared for, Fräulein Matthild', and — I say it in all sincerity — a living monument to your skill in the culinary art. It would ill become me to point out that he is digging his grave with his teeth. We do not want a nightingale the size of Josef. We prefer the appearance of delicacy, even of fragility. Unless our nightingale is to become a Wagnerian screech owl.'

'Heaven forbid,' muttered Josef fearfully.

There was a slight softening in the eyes that the Doctor cocked in the direction of the fuming Matthild'.

'You do not deceive me with your soft look, Herr Doktor,' she scratched. 'You are pleased enough to accommodate yourself at my table, you and my brother and the rest of your hungry friends and acquaintances.'

The Doctor nodded.

'With good reason, *gnädige Fräulein*. It is an excellent table. There can be no one to equal your skill in the whole of Strelsau.'

He bowed in her direction.

Matthild' sniffed. She could recognise flattery as well as sarcasm at rather less than ten paces.

'You will forgive me, Fräulein, if I remind you that we are speaking of an invalid.'

She allowed herself to be mollified, but she had still her own ideas on the subject. How could a mere man, doctor or not, estimate the intelligence and skill of a good cook? Women were tending the sick before doctors were invented. Where, she asked herself, did doctors get their ideas from, if not from women?

Knowing Matthild', and perceptive of the trend of her thoughts, the Doctor smiled, and shook his head.

He said nothing more to them, since he considered that it would be quite useless. In the end he could use his authority as a doctor. In the morning they would perhaps all be a little more sensible.

In the morning Isabella's eyes opened again. But while they stayed open, and even moved, there was no sign of recognition. The effect was of overwhelming weariness. She did not speak, nor did she move again.

Josef was quite dispirited.

'What can we do?' he said, in despair.

'Leave well alone,' Matthild' told him. 'She is after all alive.'

'Leave well alone,' said the Doctor shortly. 'Mother Nature is still the best physician. The pulse is normal, if weak. The respiration is normal. The patient is sleeping. These are good enough signs at the moment.'

They felt almost happy again at his words.

Day by day the routine of Josef's comfortable house began to assume more of its normal aspect, although the focal point was still the sick-room. The days grew shorter, but within the house it might have been Spring again, instead of the Feast of St. Stephen. It was cleaned and burnished from top to bottom by a tireless Matthild' and the servant Maria, who both found a tremendous relief in the physical effort required. The windows had never sparkled so brightly; the furniture had never been so polished; the pictures had never hung so straight. Never had such appetizing food been prepared for Josef's enjoyment.

As some colour returned to Isabella's cheeks, and her eyes were somewhat brighter, the transformation seemed to effect them all. Papa Smith went off to his rehearsals and performances at the Opera House, carrying his violin in its battered black case under his arm, and played as though inspired. His colleagues were moved to protest that such fine playing only served to expose their own deficiencies. They begged him to play as they did — routinely, and at so many bars to the Schilling. When he looked at them uncomprehendingly, they clapped him on the shoulder, and declared they were only joking.

'How is Isabella?' they would ask him.

'Is Isabella any better today?'

Most of the members of the orchestra knew her by sight, and her story, so far as he had permitted himself to tell it, was not unknown to them. They were of the opinion that Herr Smith had been very generous in taking her under his wing, so to say. When it came down to it, they did not like to think of children suffering through neglect. He was touched by these expressions of interest. People were quite thoughtful.

'Isabella is making good progress,' he would inform them, wishing in his heart that he had better news. 'Today she took a *consommé* — two cupfuls. Imagine. Fräulein Matthild' is an excellent nurse.'

They all knew Fräulein Matthild'.

They brought little gifts for Isabella which they handed to him shyly. 'From my wife,' they would say, 'for Isabella.'

Some of these gifts were entirely suitable to a child half her age, and quite inappropriate for one who was more mentally mature than their donors.

Josef himself was perhaps the worst offender in this respect. Returning home from his daily teaching visits he would invariably carry with him some juvenile toy, or a box of sweetmeats, or a packet of sugar biscuits. Occasionally he purchased these himself, for he held firmly to the belief that the way to a child's heart was made easier by means of such frivolities. Frequently they were given to him by the ladies of the houses he visited to teach his other song birds.

'For your little nightingale,' they would say as they put the gift into his hands. They were quite sincere, for the news had already spread into the great houses of the city that at last Herr Kapellmeister Josef Weingartner had found the voice he had been seeking all his professional life. They would have had every right to have been jealous, since he rarely heartened *their* daughters with any compliment other than: *Satisfactory, Fräulein Obergurgl*; or, *Our voice is not so . . . resolute as it was last week, Fräulein Schumaker*. One or two, it was true, considered that he was making a great mountain out of a very small molehill. Nevertheless, he was so happy that he now frequently shocked his pupils — and their mamas — by praising their laboured efforts to produce a noise that could be said to resemble singing. They simpered, they curtseyed, and continued to hoot and screech. Herr Josef Weingartner smiled wanly, and thought of Isabella.

As he trudged through the snow from one screech to another, he smiled on everyone he passed, and bade them *Grüss Gott*, whether he was acquainted with them or not.

At this time Christopher knew no one to whom he could communicate his joy. He was a yet a stranger in a strange land, with the little German as taught in an English Public School at his command. Driven into solitariness for the greater part of every day, sustained by the thought that it was he she had recognised and spoken to first, and that he loved her, he translated his confused daydreams into action. He took up his violin and played so long and so assiduously that all his neighbours put their fingers in their ears. It was too much, they said. Violins all day, and cats all night. Oh, it was too much. He had, they said, a wonderful virtuosity, *aber wunderschön*, and begged him to go and play in the woods. Christopher understood very little of what they said. He smiled, thanked them politely in English for their interest, and continued to play.

He remembers, with his lips twisted in a wry smile, that he was quite determined to be a worthy companion in music to Isabella when she sang again. Together they would make such music. She must sing again. Why should she not? Staring at his music, he willed it to happen.

'*Sing, Isabella,*' he cried aloud in anguish. '*Please, please sing again.*'

But singing is not like selling candles, or bread, or joints of meat. '*I only sing when I feel like it.*'

One must feel, to sing.

'I only sing when I feel like it . . . .'

He sawed away at his violin interminably. He would inspire her to sing. He would surround her with such an atmosphere of love and music that she could not help but sing again.

Day after day he pursued scales and arpeggios up and down the strings; day after day he wrestled with the most boringly technical pieces simply to acquire to perfection the mechanics of this most difficult art. His fingers developed callouses; his arms ached without relief. He continued to drive himself, and though his technique became more and more competent, inspiration was reluctant to declare itself. If anyone could have pointed this out to him he would perhaps have been spared a great deal of anguish.

In any case, as Fräulein Weingartner had remarked with some asperity, when, lost to all sense of time while practising, he had failed to arrive in time for dinner, one genius in the family is enough. Saddled with an increased, if temporary family, she had devised her own way of dealing with the situation. A pretended, judiciously-used irritability was her chief method. While the storm lasted it was fierce and awesome, and order was restored in the ranks in the minimum of time. She would chuckle to herself, and the offenders were forgiven. The memory of the running tongue remained.

'She has a tongue like a run of double crotchets,' declared Josef when he chanced to be under fire. No one knew what he meant by this remark, but it seemed to ease his feelings.

'Poor Fräulein Weingartner,' Christopher reflects; devoted to her brother; burdened with the Smith family; nurse to Isabella. Shouldering every burden. But still there had been no dinner for him, only a good run of 'double crotchets'. Afterwards in the kitchen, he had been regaled with every kind of titbit, and a cup of exquisite coffee. Matthild''s spirit was as large as her body was small.

'There now, never be late again, young man,' she admonished him. 'Practising is practising, but mealtimes are mealtimes. You are not old enough to be temperamental, to come and go as you please. In any case, one genius in the family is enough.'

*

Matthild' sits now, as of right, in the stage box, her back straight as ever, her face only a little distance above the counter of the box. She shows no reaction to the beauty of sight and sound; no emotion softens her withered face. She watches intently, nevertheless, the movements of the white-clad figure of Cordelia about the stage, and listens with reverence to the beloved voice. So long ago since she has heard it. A tear trembles in the corners of her old eyes. The Doctor leans to take her hand.

She turns to him, slowly, in the dignified way of old age.

'Does she not sing beautifully, Matthild'?'

'Isabella sings very well,' she replies, with the frail precision of one greatly advanced in years. 'It is a pity that we do not hear more of her.'

'She is greatly in demand everywhere, Matthild'. There is no one sings like Isabella. She is, in fact, world-famous.'

Matthild' nods her white head.

'I told Josef she would make something of herself. I always said so. Poor Josef. He thought he had trained her voice, taught her to sing. Perhaps he had. Who knows? He wished everyone to know that he had taught her to sing. That was his only ambition. When my little nightingale goes out into the world, he would say, she will tell everyone that her singing master was the Herr Kapellmeister Josef Weingartner, Court Secretary to the Royal Opera, Strelsau. "I shall be famous, Matthild'", he used to say. "My name will go down in the history of music as the teacher of the world's greatest lyrico-dramatic soprano."'

'Such a harmless ambition,' protests the Doctor. 'One understands it very well.'

'She has a perfect voice,' Matthild' continues, as though he had not spoken. 'I told Josef many times. She would have a perfect voice even if you had never heard of her. She has a hard heart, I told him. She is proud. She knows her own mind. She will go a long way.'

The Doctor is compelled to acknowledge the pride. About the hardness of the heart he is not so convinced. We are what others make us, and the sad stresses of life put a terrible strain on the best of souls.

'Poor Josef,' the Doctor thinks. 'If only he could have lived to savour the consummation of his ambition.'

'Josef would not have it that she was proud. He was of the opinion that she was frightened,' Matthild' continues. 'A foolish man, in many ways, was my brother. The child could do no wrong in his eyes. Beautiful, he would say, whatever she sang. Magnificent, when she was angry. What a *Semiramide* she will sing. What a . . . with the twist of his mouth I knew so well, what an *Isolde*!'

A sigh trembles on Matthild''s lips, a regret for the simplicity of men in general and musicians in particular.

The Doctor takes Matthild's hand again, and holds it with loving pressure.

'Yet part of his dream has at least been realised. He knew quite well what he was about.'

Matthild' returns the gentle pressure.

'He knew, quite well,' she murmurs. 'But he did not know what she was about.'

'Nor did she,' the Doctor reflects, but thinks it best to remain silent.

No-one could have taken greater care of Isabella than did Matthild'. They might have been mother and daughter, with a loving bond between them. If Matthild' had thought about it, she would have said: 'Here is a sick child. I must take care of her.'

Such was the care that Isabella gathered strength from day to day. Long before the new year was well advanced she was able to sit up in bed for several hours a day. She still became rather tired, however, and remained almost totally silent. She seemed quite content to rest her eyes upon the faces of her 'family' as they sat and talked about her bed. Sometimes she recognised them, and smiled as they entered the room; at other times she appeared to be puzzled, distressed at their approach, or her eyes remained blank.

Fortunately these periods grew rarer. Winter began to pass into spring; the sun became warmer. Isabella began to take an interest in what went on around her. The Doctor, gazing benignly upon her, was of the opinion that she could now sit for a time, well wrapped up of course, in the great chair by the window. He was quite satisfied with her progress so far, and such a course might help towards greater progress. But, and he fixed them with a stern eye, there must be no excitement, no talk of singing. He produced from his bag a bottle of the coloured liquid for which they had waited so long. *Three times per day, and the mixture must be well shaken.* When it was finished there would possibly be another, of a different colour. Variety is the spice of life, in medicine as in other things.

'Indeed, indeed,' Josef nodded.

Isabella was lifted from her bed and put into the great chair. Here she sat like a queen — 'A snow queen,' said Josef, 'but a very small snow queen.' The feebleness of his humour disguising the great relief in his heart — while they all fussed about her and petted her. Already she was beginning to look well again, they agreed, and quite pretty. Maria spent many hours brushing her long hair, which seemed to have grown to an inordinate length. Isabella received all their attentions in silence, but smiled occasionally.

Each day she gazed fixedly across the Square, past the indelicate rump of the Grand Duke Stanislaus's horse, past the snow-covered naked trees that lined the square, at the foreshortened facade of the Opera House. From time to time one or other of them tiptoed to the door, looked in, and tiptoed away again, satisfied that all seemed well.

One morning Maria, bearing a bowl of hot chocolate, was surprised to find her arm grasped as in a vice. She cried out at the sudden pain and almost dropped the bowl.

'Fräulein, Fräulein Isabella . . . .'

She was frightened out of her country wits. She looked fearfully in the direction of the door, and licked her dry lips.

'Maria,' whispered Isabella, 'I will walk.'

'Fräulein, it is impossible. Suppose you should fall. Suppose the Fräulein Matthild' should come. Suppose . . . . *Ach, Du lieber Gott.* That this should happen to me. Fräulein, I beg of you . . . .'

'I will walk,' repeated Isabella, 'and you shall help me.'

She half-rose from her chair, still grasping Maria's arm. Together the two held each other.

'Fräulein,' begged Maria, 'have pity on me. I shall be dismissed. I shall be sent back in disgrace to my parents. You do not know the Fräulein Matthild' as I do.'

'You do not know me, Maria,' Isabella retorted fiercely. 'We shall know each other much better yet. You shall not be dismissed. That I promise you. Yet I will walk, and you shall help me. Hold me, girl. Support me. I shall not break. Hold me, I say.'

With a monstrous effort she rose to her feet. Straight and slim she stood, the cover falling about her feet, her hair about her shoulders. The blood rose to her cheeks; her nostrils dilated with the effort of breathing. For a moment she remained, her arm about Maria's shoulders.

She breathed deeply.

'Now,' she said, 'I will walk.'

Together the two began their fearful journey about the room, one supporting, the other supported; one trembling step at a time, through a forest of almost irresistible bedroom furniture. Compelled to this great folly, Maria's breath came gasping and sobbing, the tears rolled down her red country cheeks. But Isabella, with set teeth and the iron grasp of the sick, held her tightly.

'Hold me, girl,' panted Isabella. 'Hold me, I say. We are at the door.' She essayed a laugh of triumph which became a paroxysm of coughing.

'Now,' she said, when it had subsided. 'One step farther . . . .'

Clutching Maria's shoulder she forced the girl on. She put out a frail arm to support herself against the doorpost, her limbs trembling with weakness and excitement.

'I have walked, Maria. At last I have walked!'

'Isabella! Maria! What in the name of God are you doing?'

Fräulein Weingartner almost screamed. Bobbing behind the Doctor, who contrived to fill the doorway with his bag and his voluminous clothing, she resembled a bewildered hen whose chicks had been too adventurous. The sound of her voice struck fresh terror into Maria. With a sudden twist the girl freed herself from Isabella's grasp, darted under the Doctor's arm, and clattered down the stairs into the kitchen, whence her sobs could be heard in great volume.

Isabella trembled more violently, but faced Matthild' and the Doctor. Half laughing, half sobbing, she pulled herself up, and stood proudly before them.

'I walked. I walked. Is it not wonderful?'

Still fluttering behind the Doctor, Matthild' saw the flash of that spirit, which would always be the mark of Isabella, and shuddered. How could a child who had been so ill summon such strength out of her weakness? In spite of her agitation Matthild' could do no other than admire the iron will in so frail a body (the Doctor put it more crudely: The body of a woman with the will of a particularly obstinate mule 'It is not so unnatural,' he said. But he smiled at Matthild' as he said it). Matthild' herself could not but accept the proud posture, the thrown-back head, the momentarily flashing eyes. Isabella had indeed returned to life.

'How foolish, Isabella,' she chided, but gently. 'If we had known you were so far advanced in health and strength, we would have helped you more. You said nothing . . . .'

'I did not want help, I wished to walk by myself.'

'But, Isabella, Maria was helping you. I must scold her. She had no right.'

Isabella's lower lip drooped so contritely and charmingly that the Doctor's heart turned over, and Matthild's frosty eyes became perceptibly softer.

'I was afraid after all,' Isabella confessed. 'I compelled Maria to help me, although she did not wish to do so. It was not her fault.'

The Doctor's eyes twinkled at this statement, but became immediately serious as she began to giggle hysterically.

'I walked. I walked is it not wonderful, Doctor. I walked.'

The Doctor studied her with hooded eyes.

'Wonderful,' he agreed dryly. 'And if you now have a relapse, that will be wonderful too.'

Isabella's eyes were suddenly fearful. With a last flash of spirit she tossed back her hair.

'I do not intend to have a relapse. I intend to be very well.'

But even as she made this brave declaration, the fire of endeavour died. Her face was suddenly drawn and pale. She put up her hand to ward off that importunate pain in her breast, and with a faint sigh fell into the Doctor's arms.

End of ACT II, Scene i

ACT II, Scene ii

The Inn of The Cat-in-Boots in the same village.

The next day.

ACT II, Scene ii: The Inn of the 'Cat-in-Boots' in the same village. The next day.

It was the result of a too-soon and too-brave an effort. There were fortunately no further mischances, but there were many unsolved problems, none of which have been solved in the intervening years.

He has had no communication with Isabella, following her career by the medium of the newspapers only, themselves particularly uninformative.

Now, as he watches her on the stage, and listens to that beloved voice, he sighs deeply. He cannot imagine her reason for condescending to appear again in this impoverished, third-rate Opera House. Ambition has surely been satisfied, if that were the original driving force, and if to be categorised as *Prima Donna Assoluta* was the original ambition. If riches were the aim, that too is no longer urgent, according to rumour. Why then has she returned to Strelsau after twenty years, a hiatus during which, it is well known, she has refused to appear there in public?

Certainly it is the Coronation of His Highness the Grand Duke, and his birthday. That is little reason, as he, the Doctor, understands Isabella, since she has sung before many crowned heads. It is common knowledge that His Highness has stated that he could not countenance an operatic presentation on this occasion unless Madame Isabella Weber sang. She was a child of Strelsau, as he was. It was most fitting that she should assist at his coronation and that she should be honoured for so doing (he had in fact already consulted the Court Jeweller in pursuit of a suitable design for the Honour).

Such a to-do there was, persuading her to accept the invitation, such a passing to and fro of correspondence, such a begging and pleading, such an amount of references to His Majesty-to-be and her loyalty to the city and land that bore her.

Then Madame Weber had dropped her bombshell: she would indeed sing at the Strelsau Opera at the time of His Highness the Grand Duke's Coronation. Her fee for so doing would be a heretofore undreamed of sum of Austrian Schillings. In addition, she insisted on singing the opera *Alma Perdita* with the cast as it had been sung in Strelsau twenty years ago. A world-wide search had to be made for these persons, and a special arrangement *vis-à-vis* contracts for the composer-conductor.

The Doctor sighs again. It is all too much for him. *Materia medica* he understood; the *Psyche* was an enigma he could not begin to solve.

In the short space of time between the closing of the curtains at the end of the Scene i and the opening of them for the beginning of Scene ii, some trick of sound from an instrument handled carelessly in the orchestra makes the Doctor imagine he has heard the tolling of the bell across the Square, and for a reason he cannot understand he shivers. He takes out his watch to assure himself that there is still some time — for what, he does not know. It is merely an automatic gesture.

It is indeed a quarter past the hour, by the same watch with which he had measured Isabella's pulse as she lay, white and still, in Matthild's best bedroom. Almost as big as a clock, it is, its floriated hand ticking off the seconds loudly and deliberately. It is the Doctor's pride and joy. My watch, he would say frequently, is as reliable as the sun and the moon. It has witnessed the coming of many, struggling and wailing, into the world; it has seen many go peacefully out of it. It has brought some comfort to many children as they listened to its tick, their ears pressed to its bold convex crystal.

A quarter past the hour.

Isabella has been singing almost continuously for the whole of the time, with the exception of the interval, one of the most difficult roles in modern music. During all that time she has been in full view of the audience, but, again except for one little slip, she has shown no sign of the anxiety which brought her to his consulting room, was it only a few short hours ago?

Inexplicably he had failed to recognise her. Misled, perhaps, by the fact that the attendant coachman wore the livery of His Highness Karl Ernst Maria Michael Frederik, Grand Duke of Ruritania today made King, the Doctor had been temporarily at a loss to account for the presence of the beautiful creature who had gracefully mounted his steps and entered his house. It was not often that Castle business came his way, although he was not unknown there, since many of his patients were connected with the musical pretenders of the city, and their disorders, real or otherwise, were in his care.

Many of them hovered about the fringes of Royalty, but here was a different situation. There was something about his visitor that suggested the Castle, although he could not have explained what it was. One did not expect the Castle to wait upon one; rather one would expect a peremptory summons . . . . But it was a dull late afternoon, and the snow was falling heavily. He had had, in addition, a full and tiring day, for people do not refrain from illness because it is the Grand Duke's Coronation Day.

A passing linkman called the hour. The flambent light of his torch cast a momentary illumination through the window over the face of his visitor, who had raised her veil. The Doctor caught the ghost of a smile. In the darkness which was rapidly enveloping them, his fingers sought his tinderbox, and with trembling hands he applied the flame to the lamp upon his writing desk.

The light revealed a tall stately woman clad in expensive furs; bronze-gold hair dressed in the fashionable mode; the steady gaze of green eyes, deep green they were; high cheek bones reminiscent of the Slav, but sadly with something a little less important than the original patrician element; the aristocratic length of neck. The total effect was beauty, dignity, and a composure which cannot be acquired nor copied.

The ghosts of memory stirred, prodding broken fragments with their cold fingers, seeking, sorting, endeavouring to re-assemble the picture they once formed.

A little smile softened the patrician face once again.

The Doctor cried out in sudden recognition.

'*Süsser Gott*. How could I be so stupid? Isabella! Isabella Weber!'

'Madame?' he added, uncertainly, half questioningly.

'As you wish,' she replied. 'I prefer ''Isabella'' from you.'

'I beg your indulgence, Madame.'

He could not now bring himself to repeat the name 'Isabella', although it had risen so spontaneously to his lips a moment ago.

'Indulgence for what?'

The voice was gentle, dignified. The voice of Isabella Weber.

'For my dullness. For my temporary misapprehension.'

'My indulgence, dear Doctor, would lie in the fact that it is a very long time ago since we last met.'

Still the voice was gentle, controlled, he thought, with an effort.

'Twenty years ago, if I remember correctly.'

'You do indeed. Twenty years is a long time.'

He detected a quality of resignation in the voice now, and he felt afraid. He placed a chair more conveniently.

'If Madame will accommodate herself,' he invited her tentatively.

She seated herself with an easy grace.

'Doctor, Doctor,' she protested. 'Such extreme politeness. We are after all friends of long standing.'

He agreed, with a nod of the head, that this was so. But the nod managed to convey also that the attitude even between friends may change. He glanced beyond the window into the street where stood the royal coach before his door, the horses stamping impatiently in the snow. A knot of curious spectators had gathered. It gossiped, gesticulated, pointed. Its shadows moved across the windows in the misty light from the coach lamps.

With a sudden gesture of exasperation he closed the curtains.

'There is fuel being added to whatever fire may already be burning,' he said bitterly.

'There is, then a fire?' Isabella asked.

'I speak metaphorically, Madame. If there is no fire, there has certainly been a great quantity of smoke recently. Never more so than on this very day.'

There is a sharp edge to the Doctor's tone. He returned to his writing desk and took a chair opposite her so arranged that his face was in shadow. By this means he was able to study hers more intently; to remark a lustre, a brilliancy, which were all too obviously the legacy of recent tears.

'So,' he thought to himself. 'She is a woman after all.'

Aloud, he continued:

'Your strange — shall I say, intransigent? — attitude in respect of today's events has raised a veritable fog of smoke.'

Isabella's eyes flashed. He recognised the signal.

'I cannot think of any reason why that should be so,' she said coldly.

He understood her to mean: *That may be so, and you may be my friend. But what business is it of yours?*

'One expects capriciousness in a *Prima Donna Assoluta*. It is part of the received legend. But it is entirely out of character in Isabella Weber. Not in twenty years have you been accused of capriciousness, Madame.'

'And what part of my attitude do you attribute to capriciousness?' she demanded.

He studied her face for some moments. In spite of the abrasions and the buffetings of the passing years, in spite of the evidence of recent tears, she was still Isabella Weber, proud, arrogant, beautiful. But not, he thought, as self-confident as she once had been.

'There is the matter of your presence here in Strelsau, in spite of your many protestations to Herr Weingartner that you would never, your own phrase, set foot in the place again.'

She remained silent.

'Having arrived here, you ignored your personal invitation to His Highness's Coronation . . . .'

At that she sat up and looked him straight in the face.

'How,' she demanded, 'how is that known?'

'These things are known, Madame. The whole city knows it. A rebuff such as that cannot remain hidden. Even the *canaille* out there,' he gestured towards the window, 'have discussed it among themselves.'

She shrank back in her chair.

'There was the little matter of the empty chair in the Cathedral this morning,' he reminded her. 'It assuredly did not go unnoticed.'

A moan escaped her lips.

'Then there is the matter of your disappearance from the city, in, of all things, a coach driven by a lackey wearing His Majesty's livery.'

'I had my reasons,' she cried, suddenly passionate.

But he went on remorselessly.

'Finally, your insistence, again having been invited, on singing *Alma Perdita*, as if there were nothing else you could sing with equal virtuosity. That has put the fox among the chickens with a vengeance.'

'*Alma Perdita* is a remarkable work by any standard,' she retorted valiantly, 'and Cordelia is my favourite role. That I should wish to sing it again, in Strelsau, is surely not surprising.'

He inclined his head.

'That is understood. But was it necessary to make the *geste* of putting the Director of the Opera to the almost insurmountable task of seeking out the others of the original cast, and of bringing them from the ends of the earth? For one night, Madame . . . .'

'For His Majesty's Coronation only the best is good enough.'

She glanced at him quickly, and looked away again, as she said this. He noted the lack of conviction in her voice. He smiled, grimly.

'With one exception, Madame — The Fräulein Lisa Weber . . . .'

She closed her eyes.

'They know my style. There was little need for rehearsal.'

She felt unutterably weary, suddenly *old*.

'I thought you, of all people,' she murmured, 'would understand.'

'I understand perfectly, Isabella.'

She opened her eyes at the sound of her name.

'However we try to hide it, the human heart is an open book, and can be read by anyone with the slightest knowledge of the human mind.'

Her eyes were wide, green.

'I may say, without wishing to seem omniscient, that I know everything. It has been my secret for a long time, and shall remain so. Now, a coffee, a glass of cognac? No, perhaps not, for the moment.'

Now her eyes were dark, staring. Two white lines appeared, one on either side of the bridge of her nose. Putting on his half-moon lenses, he peered through them, making a mental note of the puckered semicircles of darkness below the lower eyelids, the slight bluish tinge in the rosiness of the corners of the lips, the sharp intake of the breath from time to time. She felt his thumb upon the pulse in her wrist, firm, but as gentle as the touch of a woman, and heard the tick of his great watch.

'I shall be well enough,' she said.

He released her hand. She heard him moving about, far away it seemed. She heard the clink of a glass, and the dry rustle of paper.

'Open your mouth,' he said, standing before her.

Like a child she opened her mouth, projecting the tongue. He shook the powder from the paper on to her tongue.

'Good. Now a drink. It is only water, and will do no harm.'

Her tongue savoured the bitterness of the granules. She made a face.

'What is the powder? It is quite unpleasant.'

'What a child you are still, Isabella. It is a sedative merely. Suffice it to say that it will help your pain for a while, and restore to you some of your wonted vitality. As to the taste, have you not learned that the things that seem designed to benefit us in life invariably leave a bitter taste behind?'

She made a moue at him.

'In any case, I can do no more at this time. Unless, that is, you describe your symptoms exactly.'

After a pause she said: 'You are well aware what the matter is, dear friend. There is little that you do not know about my health, even after all these years. The condition does not improve greatly with age.'

'Age! You are no great age yet, Isabella.'

'Age is not only a matter of the number of years. You should know that better than I.'

She began to shiver, although the room was warm.

'You may have a glass of cognac now, if you wish.'

'I prefer coffee. And that I shall have from my friend Hans presently.'

'You are still stubborn, Isabella. Do you realise that you are also perspiring? I grant you that that is a reaction of the nerves. But I do not wish you to get a cold.'

'If I got a cold this moment, Doctor, it could not incommode me for at least three days. By that time it will all be over.'

He stared at her. His mouth fell open.

'I do not understand,' he said at last. 'It will all be over! What in the name of God do you mean?'

'Tonight I sing. In honour of His Majesty. After tonight I sing no more.'

Even now, he thought, she could not resist a touch of the former Isabella. He sighed deeply.

'I am tired of the endless travelling, the dingy hotel rooms, the dismal backstages. I am tired of the fawnings, the jealousies, the bargaining, the endless bickering. I am tired of women who hate and smile; I am tired of men who think an opera singer is a plaything. For twenty years and more everyone has had pleasure from my singing except myself. And there has been no place I could call home.'

She demanded, passionately: 'Have I not done my duty in that sphere of life into which it pleased God to call me?'

The Doctor remained silent. The storm must blow itself out.

'Tonight I shall sing *Cordelia*. Tomorrow I shall become, and remain for ever, Isabella Weber. I wish to live my own life, a private life, in a home of my own, with my . . . with Lisa. We will live like birds in a tree, and sing to each other.'

The storm subsided. She ceased to tremble. When she continued she was quite calm.

'Tonight, dear friend, I shall say farewell to Opera. Tonight I shall have sung my last aria. I am happy to be able to announce it while I can still sing. I shall leave in good voice, so that at least they will have pleasant memories of me.'

'It is to be hoped that you will reconsider the matter,' he said slowly and, he felt, rather stupidly.

'I shall not reconsider it, dear Doctor. I am quite decided.'

'Why, Isabella, have you come to Strelsau to demonstrate what I can only call a quality of exhibitionism, of bravura, which is really quite foreign to your character?

An expression passed over her face like the shadow of a cloud in April.

'One has to have a last performance somewhere,' she said, somewhat defensively. 'This is my birthplace; I began to sing here, I shall finish my song here.'

There was nothing he could say. He was concerned with the well being of the body, not the *capriccios* of the mind. It was a pity that such a voice should be so prematurely lost to the world, if she were indeed serious. It was a pity that there would remain only the memory of it, growing fainter, until all knowledge of it was gone, and only words remained.

'Ah, they would say in times to come, but you should have heard Isabella Weber.'

All Papa Smith's dreams; all his (the Doctor's) medical skill; all Josef's careful and loving instruction; all Matthild''s devoted attention in sickness, unhindered by personal duties and demands, had been given for this, that all must dissolve and vanish away. The name of Isabella Weber would remain for a short time on everybody's lips, but the memory of it would fade, until it became a mere entry in the wash of musical histories and dictionaries.

'*Maestro*,' future students of the art would ask their professor. 'Tell us of Isabella Weber. Who and what was she?'

And the *maestro* would take down his dusty volume and read, peering through his spectacles at the small print, the inscription on the paper tombstone:

WEBER, Isabella. Lyrico-dramatic soprano. b. Strelsau 1856. Debut: Vienna (*Fidelio*), at the age of twenty. A pupil of Weingartner and Marchesi. Appeared in opera at Strelsau, Dresden, Leipzig, Munich, Berlin, London, Milan, Rome, etc. 'Her voice was an instrument so unique that its like has not been heard in our time'.

'There, my children,' the *maestro* would tell them, closing the volume with a faint bang of finality. 'If you practise your scales sufficiently, and study your theory of voice production diligently, you too may one day sing as Isabella Weber sang. At two hundred pounds per performance, no less, and increasing. Your names too may be on everyone's lips.'

'But, *maestro*, what you have read tells us nothing. *How* did she sing? What made her so different from the others?'

How indeed was she different? Was her voice the sound of the cold ripple of winter water over pebbles? Was it the murmur of the wind through leafless branches? Had it the precise passion of a soaring lark or the warm tones of the voice of a loving heart? All these certainly issued from an instrument so unique that at its sound the bowels melted and the heart was filled with love and compassion.

How could the old *maestro* explain to his ambitious students what made Isabella Weber so different?

God had heard her sing, and knew that it was good.

He was recalled by the sound of her voice.

'Madame? Forgive me. I . . . .'

'God has heard me sing. Is not that what you said? It is enough.'

'That is what I said, I recall.'

With an effort he remembered his obligation.

'Are you able to sing your part this evening?'

'I must. What is to be will be. At this moment I feel very well. Your powder is most efficacious.'

She rose from her chair, and stood before him.

'In the event,' she continued, ignoring his searching scrutiny, 'I was not seeking medical advice. At least, not directly. I came to ask a favour.'

'A request from Madame Weber is my command.'

She smiled.

'It is merely that you should attend the Opera this evening.'

'The request is superfluous. I am to be there by invitation.'

'I understand so. That is why I have taken the liberty of making other arrangements for you.'

He bowed, slightly, ironically.

'In place of the seat you were to occupy, you will take the stage box. You will oblige me by giving the pleasure of your company to Fräulein Weingartner . . . .'

'Frau Doctor.'

'Frau . . .?'

'My wife. We were married shortly after Josef died.'

'I had not known.'

The colour rose to her cheeks in confusion.

'It will be pleasant to meet Matthild' again . . . and, of course, any other friends . . . .'

She paused at the shadow which passed across his face at this echo from the past.

'You are angry. You think me impertinent.'

'Indeed not. How could that be? There are times in life when we must state our feeling, directly or obliquely, or be misunderstood for ever.'

But clearly she was put out in her turn.

'That box, as you know, has a door which communicates directly with the "Bridge of Sighs". This evening . . . .'

'This evening?'

'Yes, Isabella?'

His voice was gentle, as though he were comforting a sick child.

'Nothing, it is nothing. I have a premonition, perhaps. It is nothing I could put into words. Perhaps I am foolish, nervous. But I wished you to be there. As a precaution, you understand.'

He nodded. At his cupboard he counted out some tablets into a small circular box, and scribbled on the cover.

'Should the necessity arise, take one of these tablets.'

Without a word she put the box into her purse.

'You are still very patient with me, and a good friend.'

'I have been your devoted admirer for many years, Isabella.'

She drew on her gloves with precision.

'Thank you, Doctor. And now, having been *in absentia* during my Prince's Coronation day, I must prepare to entertain him, once more, this evening. Till then, my good friend, *auf wiedersehen*.'

'Till this evening, Isabella. Go with God.'

At the carriage she said to the lackey, as he stamped his frozen feet: 'To the Opera. The stage door. Thereafter you may return to the Castle, to tell what tales you will.'

*

The apparent relapse was not so serious after all. It was merely the consequence of too great a first effort. Before long Isabella was once more royally installed in the great chair by the window, savouring the strengthening sunshine, observing the gradually emerging rump of the Grand Duke Stanislaus's horse from its covering of snow, noting the increasing number of the good burghers, of Strelsau as they crossed and recrossed the Square about their daily business.

Each day she walked, a few steps at a time, about the room, helped and supported by Maria. A great bond grew between these two, which was to last many years; the serving and the served, but with no taint of servility. Such exercise inevitably took a great toll of her weakened body, and invariably she regained her chair flushed and trembling, but full of the pride of achievement. If the pain, happily somewhat diminished now, refused to give up its grip entirely, she told no one, except that the occasional involuntary gesture informed the Doctor that all was still not yet well.

In fact, she spoke very little at all. Only Maria received any confidences she might have made, as secret as the Confessional.

Her visitors, who came in increasing numbers, found her a strange little creature, so cold a response did their fawnings and condescensions have on her. Despising them, Isabella had to accept their false affability, since Herr Weingartner had instructed her that both he and she depended greatly on their goodwill. They had, as it were, shares in the property, and they must be made to feel important.

'It was going on long before you were born, and it will continue. Beethoven, Haydn, Mozart . . . . Artists are at the mercy of their patrons. You may behave as if you are a queen, but you will be a queen by their favour only.'

But he was disquieted by her stony look and curled lip.

'What a strange little girl,' the ladies of Strelsau exclaimed to their acquaintances later. 'So sulky, so sullen.'

There was no effusive humility in her; they received no fulsome thanks for condescending to visit her during the weakness of convalescence. There was really nothing about the child that seemed to be worth their effort. She had scarcely opened her mouth. They were not at all sure that 'Dear Herr Weingartner' was not imposing on them. It was all very difficult, they sighed.

'It is to be hoped,' they said to Herr Weingartner in the manner of a veiled monition rather than the expectancy of value for money, 'that she can really sing. There seems to be little else of promise.'

'Isabella has been very ill, Baroness,' (or Countess, or Frau Postmaster-General. It was all one to Josef who they were so long as they were sufficiently wealthy and influential).

'Such a pity,' they murmured. 'The child must have been quite pretty before she became so thin. There is nothing like good basic nourishment . . . .'

Matthild' bridled, and her pale face stained with colour.

'I assure you,' she began.

'. . . And what a disturbance for the household. Fräulein Weingartner was really the most unselfish person,' they said. 'An Angel, especially when it was so unnecessary — well, really, was that not so? — for her to put herself about at all. There were so many sick children; if one tried to care for all of them . . . . There were others to do that . . . .'

'Sick children there may be, the more's the pity. There is only one Isabella,' Josef informed them, as disarmingly as he could.

They smiled.

'So droll, Herr Weingartner, so *komische*. We believe he is half in love with his nightingale. At his age, too.'

They tapped his arm playfully with their fans, and he closed his nostrils against their perfume. He smiled at them, and bowed.

One could only hope, they continued to say, that the child would be grateful enough. For their part they thought that ingratitude was one of the so very disagreeable qualities of children. Their smiles faded a little as they inspected Isabella again. They spoke about her in loud voices, as if she were deaf, or an idiot. It was indeed a pity that she was such a poor little thing. So thin, with such unfortunate green eyes and that peculiar hair. The nose too was perhaps a little . . . out of proportion, was it not? One could not really imagine that such a child would ever sing. Herr Weingartner was always about to discover a new nightingale, only to find — fortunately before much time, effort and money had been thrown away — that it was merely a backyard hen. No doubt his Isabella would turn out to be a bird of the same feather. Their own daughters were doing splendidly, so splendidly, indeed, that they — the respective Mamas — were sometimes quite frightened at the splendidness of it. One sang, of course, when one was a younger woman, but Griselda (or Anna or Gretchen) is so much better, thanks to dear Herr Weingartner. Perhaps there was the little difficulty of accurate singing on the top notes, but that was never really noticed, especially if the young ladies were of good family.

Herr Weingartner, they — the Mamas — said, was a quite remarkable teacher.

The daily visits became rather fewer after a time. So disappointing, they said. There were no remarkable demonstrations of the possibilities of the human voice; there was no music, no singing. Herr Weingartner's nightingale had absolutely none of the conversational qualities they might have expected from his sustained eulogies concerning her.

'A quaint child,' was the most appreciative comment that they were able to pass on to their *confidantes*.

Certainly Isabella remained strangely docile, even amenable. At Josef's tentative inquiry, and with the Doctor's grudging permission, she began her simple exercises again. Firstly, she sang from her chaise-longue, intoning rather than singing, since Herr Weingartner was still fearful lest she should have suffered some injury to the voice. Then, feeling stronger she sang from the chair, supported by Fräulein Weingartner's best cushions. Finally, she stood once more by the *forte-piano*, her thin body illuminated by the weak winter sunshine, her hair a tumbled mass of bronze-gold. The slanting light turned everything it fell upon to gold — the frame of the picture of *The Music Lesson* (by an unknown Master, it later transpired); the shining copper bowls void for the moment of fruit; the ancient but well-polished furniture.

Isabella breathed deeply, happily, recalled as it were to life. The golden notes, weak but unimpaired, filled the room and the house. The singing birds were silent. Matthild', relieved of her recent onerous duties, was intent upon her embroidery. In the kitchen Maria paused in her baking, her young heart filled with thankfulness, and raised her eyes heavenward. '*Gott sei dank*' . . . .' she murmured.

Once again passers-by in the street paused in their perambulations to listen. Standing beneath the window they dared scarcely breathe lest they miss even one note. Isabella sang scales and runs from Signor Rossini's workshop book; to her small audience in the street they were a stairway to Heaven.

'Our nightingale sings again,' they whispered, nudging each other in glee, for the voice was quite unmistakable. For half an hour they stood there, until the appearance of Fräulein Weingartner behind the curtain warned them to be about their business.

They were in no way abashed.

'The Fräulein is happy,' they said as they went their ways. They were well enough acquainted with that dour face to read all its expressions.

Isabella grew stronger in health, and her voice began to recover its flexibility. She practised simple songs and sophisticated airs; she sang arias. Time passed unremarked. The great bell of St. Michael continued to toll for the dead hours, and the vibrations spread the ashes of the dead time into every corner of the Stanislausplatz. The dusty dolorous sound crept into Herr Weingartner's *salon*, where it was unheeded; it could not prevail against the living beauty of Isabella's voice.

The Doctor remembers with distaste Josef's reaction to the situation. His pleasure was wonderful to see. He smiled, he beamed. He hummed snatches of song with Isabella in what, for a singing master, was an execrable voice.

'You are indeed a nightingale, Isabella. Your voice is truly recovered. My teaching is good, *nicht wahr, Herr Doktor? Wunderschön, aber sehr wunderschön.*'

He raised his voice.

'Matthild',' he called, 'Matthild', *komm' gleich* . . . ah, there you are. A glass of your cherry wine for the Herr Doktor and yourself, dear sister and a *Stein* of beer for the Herr Kapellmeister Josef Sigismundus Weingartner, inventor of the New Weingartner Method of the Art of Singing.'

'Are you mad, Josef?'

'Indeed not, Matthild'. On the contrary, I am quite sane. Our Isabella sings divinely, in spite of everything, thus proving what a wonderful teacher I am. Today, Sister, Doctor, you are witness to the birth of a new phenomenon in the world of music. Isabella is the first to benefit by it. We shall make history, she and I.'

'That is as may be, Josef. Those who live longest see the most.' But she poured the wine nevertheless.

Josef laughed until the tears ran down his cheeks, and his canary-coloured waistcoat, rather dimmed by the winter's tobacco ash, heaved tremendously.

'The Weingartner Method . . . . That is good.'

He choked and spluttered between gusts of laughter and draughts of beer.

'That will make Marchesi look to her laurels.'

Clearly Josef was happy. It is sad, the Doctor thinks, that he did not live to enjoy the consummation of the Weingartner Method, or the Marchesi Method, or whatever Method it was that produced the voice I am hearing this evening. Certainly Josef knew his subject; his lack was the understanding of the human mind.

'I said that she had a hard heart. I told Josef so many times. He would not believe it.'

Recalled from the past, the Doctor is aware that Matthild' is indulging in her favourite exercise of reading his thoughts. He puffs out his lips perversely.

'She too has her memories,' Matthild' continues, 'and they are beginning to beset her.'

The Doctor continues to puff out his lips, until the force of what Matthild' has said strikes him. He turns to her in surprise.

'She has made peace with her conscience at last,' Matthild' murmurs.

'So much is obvious.'

'My dear Matthild', in what way is it obvious?'

'Do you not hear her singing?'

'She is singing perfectly, beautifully.'

'That is precisely what I mean. After all these years we are at last hearing what Josef intended we should hear. The perfect voice, untrammelled by earthly cares. Sometime during the last twelve hours, one feels, a great decision has been made.'

The Doctor is astonished and bewildered at Matthild''s percipience.

After tonight I sing no more. Was it not Isabella who herself had said those words only a matter of three hours earlier?

'I think, my dear,' Matthild' said softly, 'you are as near to Heaven, musically speaking, at this moment, as you are ever like to get.'

Again he turns startled eyes upon his wife. Her face is impassive. She gives every impression of being uninterested. After a pause, she continues: 'Men's eyes are so much on the stars that they rarely see what is going on beneath their noses.'

'Matthild' . . . .'

'Hush. She sings . . . .'

His mind is disquieted. What does she mean, and to whom could she be referring?

He reviews the situation in his mind as Isabella sings.

She could be thinking of Herr Smith, who had indeed picked the child almost literally out of the gutter, fed her, clothed her, treated her for the first time as a human being. He had fired in her the innocent ambition to be a great singer; for himself he had dreamed of a cottage in the green fields of his native country, with apple trees in blossom, and a lark soaring. When these delights came too late, he charged her with indifference, worse, with ingratitude.

Matthild' could be perhaps thinking of Christopher, that easy-going young man who was even now standing in the wings, his lack of character redeemed only by his love for Isabella. What dreams he may have had of being one day a great violinist have long since vanished. Now he carries his twisted hand in a black velvet glove, a permanent reminder of Isabella's once bad temper and carelessness, albeit under great stress. *His* love? For Isabella. The Doctor is startled by the inappropriateness of his own thoughts. Love?

'What have I to do with love?' Isabella demanded of the image staring at her from beyond the dark mirror. So many ghosts lurk there; the ghosts of pretended love; the ghosts of substitute love; the ghosts of unhappy love.

They peer at her from the dark corners of the room beyond the glass. They shriek with silent laughter.

'What has she to do with love?' They mouth silently to one another. 'Why should she need love? She has a wonderful voice, she has riches beyond reckoning, she is admired by all the world. But she is Isabella Weber, the proud Isabella Weber, and she has no soul. To her, love is a foolish thing.'

She began to be afraid. Has she indeed no soul? Had the pursuit of perfection denied her the ultimate gift of God to all his children? She tried to imagine herself with a soul, and without a soul. She could perceive no difference.

Was there a meaning to the word 'love' which she did not, could not, comprehend?

'Indeed,' Christopher had said, some time after she had recovered from her illness (so long ago), 'Love is a foolish thing. It is fantasy; it is a mirage compounded of pride, humility, selfishness, generosity, self-pity, confidence, jealousy, tolerance. Its bonds are as strong as iron, and quite easily broken. Love is terror and ecstasy, weakness and strength. When a man loves he is at once a child and more than a man; when a woman loves she is the All-Being. When we are not in love we see how stupid it is. When we are in love its agony is the most glorious feeling we can ever experience.'

At that time Isabella believed all that Christopher had said. She did not question either his philosophy or his sincerity. But after this feat of eloquence he was embarrassed, for all he had wanted to say was that he loved her; he perceived that she was now afraid of his intensity. Her nervous smile reflected her inability to understand the force she only vaguely realised that she had awakened.

'You put a great burden on one little word,' she said, hoping that by a little raillery she would be able to dissipate the tenseness of the moment. She refused to see the pain in his eyes; she could not guess the sudden desolation in his heart.

'You are hard, Isabella. You are . . . .'

'Yes, Christopher,' she said softly, 'what am I?'

But he could not tell her that he had placed her upon a pedestal from which she had fallen.

'Isabella, I love you. There, it is said.'

Isabella poked at the grass with the point of her parasol, a dainty object of silk and lace, quite the fashion among the ladies.

'Am I not your sister, Christopher?' she reflected.

'By adoption only.'

'We have spent some years together. Since that is so, there would seem to be no special virtue in this thing you call love. But since men talk so much of it, and since women seem to want to hear them talk of it, there must be some virtue in it. Why do you love me, Christopher? Of course you love me, just as I love you. Why not? But at this moment I do not feel any of those things you have spoken of.'

'It is not possible to explain further, Isabella.' He bent his head, and looked at nothing at all. The first words she had ever spoken to him were sharp in his memory:

Why are you dreaming, boy?

'I have loved you since the day I first saw you. The day I first heard you sing . . . .'

Isabella smiled.

'How absurd, Christopher.'

He threw a stone angrily into the lake. It was followed by more stones. She realised that she had said quite the wrong thing, and hastened to make amends.

'We were both children, Christopher, and strangers. And if you loved me because I sang, why, you have heard me sing every day, and you will hear me sing a great many times more. That is no reason for this love you talk of.'

'Perhaps you will understand one day. And when you do understand it may be that you will sing even better than you do now.'

She shook her head.

'That I cannot understand,' she replied, angry in her turn. 'I sing always as well as I am able. It is a trick of the voice, that is all. I cannot think of any other way, nor of any other reason. In any case, there is not much else I care about.'

'How can you say that, Isabella? Is there nothing else you want of life?'

'I do not know. I have never thought about it. I do not think that there is anything in life I want more than to sing. I must sing, although I do not know why. I thought it was to be such a simple matter.'

Before they left the place, with the sigh of the wind among the trees, and the malicious giggle of the stream over the pebbles, she allowed him to put his arms about her and to kiss her after his fashion.

There was, she remembers, no satisfaction in it for her.

The Doctor sighs. Indeed Isabella has her memories. But Matthild' was not referring to either Papa Smith or to his son. The Doctor glanced up and across the darkened auditorium. There was the key to Matthild's insight.

Fräulein Isabella Weber in Alma Perdita, *at the Royal Opera House, Strelsau, on the occasion of the Coming-of-Age of His Highness Karl Ernst Maria Michael Frederik, Hereditary Grand Duke of Ruritania, etc.*

It was Isabella's coming-of-age too, and in recognition of that achievement and to celebrate her return to Strelsau as an established singer from Vienna, Herr Weingartner had given a little dinner party, at which the Doctor was a guest. The details of that evening, and the subsequent events, are sharp in his memory.

Admired to the point of foolishness by her menfolk, and as happy as a child, Isabella sang her role that evening divinely. The Grand Duke, sitting in almost solitary state, was extremely diverted by this new singing bird, and his birthday — which until then had been without specific incident — acquired a new interest. He irrationally added himself to the list of those to whom Matthild''s cryptic remarks might have reference.

Newly-crowned King, he sits in the Royal Box, as he sat so many years ago, but not in such solitary state as then, nor with such freedom of spirit. His shoulders are bowed with the weight of years and the newly-imposed burdens of State, inadequately supported by a wife who was none of his seeking and for whom he long ago acquired a sharp distaste. The Order of the Red Rose of Ruritania glitters upon his chest, symbol of autocracy, aristocracy and propriety. He sits unmoving, unsmiling, a Prince who might be considered to have experienced the fulfilment of his desires. His eyes are fixed at that part of the stage which is occupied by Isabella, and the Doctor, skilled in reading faces, perceives that by whatever steps His Majesty has arrived at his present situation, he has regretted them many times over. He too has his memories.

How beautiful you are, Fräulein Weber . . . .

You and I shall celebrate more intimately our double anniversary . . . .

Not that the Doctor — nor anyone else for that matter — was privy to these tender exchanges, but he was wise in the ways of men.

Again he consults his watch, startled to realise how many thoughts can run through the head of an old man in only a few minutes, and how slowly time passes when there is fear and uncertainty in the air. Isabella is singing beautifully, with no sign of distress, but he has a deep feeling that all is not well. He has given his opinion to Christopher . . . yet it is scarcely that young man's fault. A *Prima Donna* is the last person in the world to whom one would presume to dictate, and of all of them Madame Isabella Weber in particular would condone no interference in her decided course of action.

There is no sign that for almost two hours she has been singing, almost without a pause, with the exception of the Intermission, one of the most gruelling roles in opera, and the worst — or best — is yet to come. The second Act of *Alma Perdita* is the longest, requiring the greatest effort on the part of the principals, particularly the soprano.

A gentle music, the most unlikely precursor of what is to come causes the group on the stage to fluctuate from pattern to pattern, arranging itself always with Cordelia as the focus.

'Hurry, hurry,' the plaintive ghosts whinny in a dressing room empty of all save memories and roses.

'Hurry,' they mew, urging on the sound of His Highness's footsteps approaching along the corridor. But the sound is as ghostly as the spirits themselves. They shriek in frustration; they withdraw gibbering in frustration into the baseless fabric beyond the looking glass. The footsteps halt; they turn away, and the sound of them dies. She has rejected his advance. She has denied his belated desire to explain . . . to excuse. From beyond the looking glass there comes the sound of malicious ghostly laughter.

The room is empty, full of the scent of roses. The sharp tinkling notes of the travelling clock are blunted in the perfumed air. The sound falls unnoticed to the bottom of the heavy perfume, there to lie silent and lifeless.

It is the half-hour.

The music is bitter-sweet. Cordelia listens attentively to the advice offered her by such of the villagers as are sorry for her and wish her well. She knows that their arguments have logic and make sense. But she is a woman. She has been degraded beyond bearing. Standing in the green and gold of the stage lighting, confronting her betrayer, her pride and scorn are terrible to see. Many times, the Doctor recollects, has he seen a corresponding expression on her face. This is not art or artifice; this is reality, the expression of an inner arrogance she still cannot conceal.

Giovanni Tortoni, world-renowned *tenore*, playing the villainous Duke, eyes Cordelia warily, and perspires at the thought of the wrath to come. He trembles in the anticipation of the fulmination of disdain, contumely and contempt that will play about his luckless head.

He voices his lines with some timidity. He begs. He pleads.

Forgive me, Cordelia . . . I beg of you.

She laughs in his face, a bitter, ironic laugh.

Only trust me. We will marry . . . . You only shall be my bride.

Andelli nurses his musicians gently, leading to Cordelia's sub-aria *Non mi tentar* — 'Do not tempt me,' which is followed almost immediately by the aria *Te amavi*, interrupted only by an agonizing cry from the Duke of '*Cordelia*'. Repeated in a minor key, it is the jewel of the work. No longer is Cordelia a lost soul; she is outraged womanhood, than which there is no greater implacable antagonist.

She has scarcely fallen to silence on the last note of the aria than the audience rises to explode its applause. The stage freezes. Andelli's baton hangs uselessly by his side. Matthild' sits quite still in her chair, not acknowledging by the slightest change of expression that she has been listening to the ultimate demonstration of the vocal art. The Doctor, fearful of the consequences of that effort, finds himself clasping and unclasping his hands, a nervous gesture quite unusual in him.

The applause continues, loud and prolonged. One or two crude souls demand a repeat.

Isabella takes a step forward. In the light it is to be seen that, apart from a disdainful flaring of the nostrils, she is in full control of the situation. She lifts her head and speaks one clear word:

'No.'

*

'But, Isabella,' began Josef.

'No,' she repeated.

She was angry; so much was apparent. Upon whom the brunt of the impending storm would fall was a matter of conjecture. That it would fall, and very soon, there was no doubt. The Doctor could only guess that Josef was in for a bad quarter of an hour. For himself, his appreciation of the situation could be measured by the degree of irony upon his remarkably bird-like features.

Matthild's needlework lay neglected in her lap, but she was calmly watchful; Josef, his wineglass drained, perambulated about the salon, pausing now and again to look musingly at Isabella. As though unaware of this attention, she sat in her chair. There seemed to be a wall about her, a hard crystalline wall.

Josef paused in his perambulations and stood before her. He rubbed his hands together, and beamed ingratiatingly.

Isabella's expression grew colder.

'Isabella, we have aroused their curiosity. We have stimulated their generosity. We have almost achieved the desired end.'

Perhaps it was the feeble attempt at cheerfulness, perhaps it was the blatant humility of it that released the tension. The slow expansion of anger spread about the room, heavily and depressingly. The wall of crystal disintegrated, vanished. Josef no longer rubbed his hands.

'*So droll, our dear Kapellmeister*,' said Isabella. '*So witty, our esteemed Herr Weingartner*.'

She mimicked with ease the originators of these remarks in both voice and in facial expression.

'One must accept these little compliments,' said Josef. 'Everyone knows that they are quite meaningless. Nevertheless they are well meant, a pleasant contribution to friendship.'

Isabella eyed him stonily. It was clear that any pleasantry, well meant or otherwise, that could be said to contribute to friendship was lost on her. She went on as if she had not heard him.

'*Perhaps the child can sing after all. There seems to be little else of promise. Herr Weingartner has such a way with children. If there is any talent at all he will bring it out. Dear Herr Weingartner* . . . ' Isabella almost sneered.

Josef's features were suddenly ashen.

'Isabella, my child . . . .' he began.

She turned upon him in fury.

'I am not a child. I am not your child. I am Isabella Weber. Do you think I do not know what is in their minds — those of them who have minds? Why must I submit to the indignity of being inspected and appraised by such people? I, Isabella Weber. I, who can sing, while their daughters can only howl like cats at night. I am thin; I am ugly. I am only a screeching backyard hen. If that is all they see, after they have given themselves the trouble of descending upon me in their patronage, they are . . . they are . . . .'

Her voice trailed away.

'You are mistaken, Isabella. You misjudge them.'

He paused, uncertain as to how he could explain the wider aspects of life to one who had so far trodden only the narrow ways of experience.

'Patronage has been an accepted way of life since the days of the Creation, and will continue to be so.'

He was gathering courage.

'It may be that you consider their knowledge of music and singing to be inadequate. It may be that their criticism is without understanding, their appreciation — at the worst — tempered by what they like. They know what they like, Isabella, and they have the means to obtain it. They can be generous to a fault.'

There was no response from Isabella. He did not know whether she understood him or not. He did not know, even, if she had heard what he had said.

'Genius is genius,' he continued, 'but it is dependent entirely upon those who are not geniuses. Genius can only take what advantage it can gain from such a situation, and ignore the rest. It is perhaps unfortunate that the unjust distribution of material wealth which enables them to patronise also encourages them in the idea that they can speak vulgarly on any subject which for the moment interests them. You may be assured that they mean no harm by their chattering. They lack indeed the wits to be other than banal.'

The Doctor was interested to perceive that Isabella's expression was changing from cold anger to sullen sulkiness. Josef could be more persuasive than he realised.

'Whatever you think at the moment, Isabella,' Josef continued softly, 'you are Isabella Weber. You are unique, as every person is unique. You will be a great singer. They do not know that of their own experience. But I know it, and because I say that it will be so they are willing, even eager, to pay for your education and training in Vienna. To that extent it may certainly be said that they have a share in you. They feel, without doubt, that they have the right to inspect their purchase.'

'No, Josef, no.' The Doctor was startled at the vehemence of his own outburst.

With a suddenness that took them all by surprise Isabella rose from her chair. She stood to her full height, her eyes blazing.

'So! Their purchase! I am to be bought then? I am to be examined? I am to be sent on exhibition to Vienna, to be paid for if satisfactory? *See what we have bought, good people of Vienna. A new singing bird from Strelsau, with an unfortunate nose*. Is that what they think of me? Is that all you think of me?'

Josef made an ineffectual gesture of capitulation. He had quite simply wished to discuss plans for Isabella's immediate future, unaware of the smouldering fires of resentment that had now burst into flame. Each of Isabella's outbursts came as a shock, more so perhaps because he could never remember how or why it had begun, nor what he had said that could produce such a situation.

Matthild', blind to the pitfalls looming ahead, came to her brother's rescue.

Her choice of words was as unfortunate as her choice of the occasion. The tone of her voice — for the first time in speaking to Isabella — was impatient.

'You are talking nonsense, Isabella.'

Isabella's reaction to this was only to be expected, was the Doctor's private opinion. She turned the full blaze of her eyes upon the hapless Matthild'.

'Nonsense, Fräulein?'

Isabella's voice was as cold as ice. Her question really permitted of no reply.

'I talk nonsense, do I, *gnädige Fräulein*? It would be of great interest to me to learn why you consider that I talk nonsense. Perhaps because I have been ill — for a reason unknown to me — so that I am at the moment incapable of distinguishing between sense and nonsense. Perhaps I am after all just a child

— *my child*, says Herr Josef. You have nursed me, and for that I am properly grateful, but I know, dear Fräulein, that behind your anxiety for me there was also the anxiety for Josef and his fear that he might at any moment lose his singing machine.'

Matthild' matched coldness with coldness.

'That is not in the least true, Isabella.'

Isabella turned upon the Doctor.

'Why did you not let me die? It would have been better.'

The Doctor shrugged his shoulders.

'Your time is not yet, Isabella. It is God's will.'

'God's will? Words, words only. I was not allowed to die because I can bring fame to Herr Josef with his ''*Weingartner Method*'', and perhaps, who knows, some additional fortune. So you attended me with the greatest care. I am indeed grateful, dear Fräulein Matthild', dear Herr Kapellmeister Josef Sigismundus Weingartner.'

The irony in her voice was not lost on any of them, but she had reduced them to silence, the more bitter because of the half-truths her accusations contained.

'Really,' thought the Doctor, 'such a child. But a child with the force of character of one far older and more experienced.'

Isabella's tone was still cold and clear. Anger was still in the ascendant. How different she was, Josef reflected, from the child who had stood before him only three — or was it perhaps four? — years ago, dressed in a black frock with a white belt, that shining hair bound in a new green riband. A child who wished to sing.

'I am supposed to be grateful too, am I not, to the selection of Josef's acquaintance who call themselves ''The Friends of Music'' who have seen fit to visit me? Rest assured they did not visit me with sympathy in their hearts. They came because it provided them with another excuse to enter your house, to drink your wine, to examine your situation in general. They came because it amused them to observe your deference to them as if they were of the greatest importance.'

'Isabella.'

Matthild's involuntary cry was almost one of agony. In these few words Isabella had almost shattered her comfortable little world. In Matthild''s world one did not ask why people were friendly; one was sufficiently grateful that one had friends.

Isabella regarded her with pitying eyes.

'They each came,' Isabella went on, 'because the others came; because they could not bear to miss anything that might provide them with gossip during the *longeurs* of their aimless afternoons. They came to see Herr Kapellmeister Josef Weingartner's latest prodigy — by invitation . . . .'

She paused at Josef's exclamation of astonishment.

'Yes, Herr Josef. You did not think I knew that, did you?'

'I thought . . . I thought such a course might help, might give you an interest,' he muttered.

Isabella ignored him.

'They are amused by his pathetic belief that one day he will discover the perfect voice; they came because by nature they are inquisitive, condescending, and patronising. They came because their minds are empty and they have no soul. They came because, although they realise that Herr Josef is using them for their money and their influence, they cannot resist *his* patronising them in his turn.'

She paused. She drew breath; there was a suspicion of a sob in her voice.

'At last,' the Doctor thought, 'the storm is over.' His heart was suddenly lighter.

But the storm was not entirely over. The thunder was losing its force, but there remained one or two belated flashes of lightning.

Again she faced Josef.

'You called me a nightingale. Is the nightingale to be bought? Does he sing because the rich burghers of Strelsau have paid him to sing?'

Josef rallied, gathering the pieces of his courage.

'Isabella, you are much more than a nightingale. You are the voice of Heaven itself. I respect your feelings and your opinions in this matter, but you must believe me. You must have faith in me.'

Now the tears gathered fast in her eyes, and Josef rejoiced.

'I have little reason to believe anyone,' she said at last. 'I am a piece of merchandise. Why should I concern myself with your bargainings?'

Her cheeks glowed with shame and anger that she should be treated so.

'These people wish to hear you, *you*, Isabella. Our little *matinée* is to be the first step in your future career. These people are — ' he hesitated on the word — 'important to you. It is to be regretted that I can do nothing more for you. For that reason you must now go to Madame Marchesi in Vienna, who will help you in ways I cannot.'

'I will go to Madame Marchesi in Vienna. I dislike to sing before your pretentious "Friends of Music".'

'If you do not sing before them, Isabella, there is little chance that you will go to Madame Marchesi.' His voice was quite sharp. 'Consider very carefully. There has never been, I swear, a voice like yours. You can sing what you will, where you will. The world lies before you.'

There was silence for some minutes.

'By kind permission of the Burghers of Strelsau, I presume.' She could not resist a last thrust. 'Very well, Herr Josef Weingartner. I will believe you. I shall sing what I will, when I will. Since the world lies before me, I will take it. I will take everything it has to offer. I will sing everywhere I am paid well enough to sing.'

Josef's sigh was as loud as a summer breeze among the leaves.

'But I will never sing in Strelsau.'

Once again Josef was left with the feeling that he had lost the thread of this horrendous discussion, that he was groping his way through a mist beyond which his goal might only be tantalisingly glimpsed before being lost again, this time perhaps for ever. He rummaged within his memory to discover what he might have said or done that had offended her so deeply; he sought to discover in what he might have failed in his dealings with this capricious musical phenomenon. He could discover no such point. He had never encountered such a difficult pupil. Yet he had a feeling that sometime during his association with Isabella there had been moments of happiness. In a calmer mood of self-criticism he blamed himself for her lack of balance. Taking over from Papa Smith in his anxiety to develop as he could her natural musical talent, and to present her to those citizens of Strelsau who were most likely to be of help in financial or in other ways, he had ignored, or perhaps forgotten, the problems attendant upon being merely human.

Talking the situation over with Papa Smith, with Matthild', with the Doctor even, Josef found that none could offer any useful advice.

The Doctor merely shrugged his shoulders.

'Leave the child alone to grow up in her own way,' was his only comment. 'She will find her own feet, and other people will have to put up with it, as they have done since the beginning of the world.'

It was with difficulty that Josef tried to follow this advice. The result, he considered, was scarcely successful. As Isabella came to realise that her voice and her singing were more than ordinary, she developed a somewhat high-handed, even insolent attitude. She was so young, and so accustomed to being the centre of her little world, that she could not understand that her attitude, though innocent, was hurtful.

Josef was compelled to admit that there was a great deal of truth in her bitter words, uttered in ignorance of 'The System'. He strove to believe that she knew, as soon as the words were spoken, she was hurting those who loved her, but that she could not bring herself to make any kind of amends except to apply herself with increased determination to her lessons. He hoped fervently that experience would teach her what he could not.

There seemed to be very little possibility of this happening in the near future. The atmosphere of Strelsau, especially with regard to music, was quite unfavourable to the development of a well-balanced character. Those would-be patrons were far more interested in reading accounts of their activities in the newspapers, or having their names in the programmes and in the Civil List. They wished their interest and generosity to be noised abroad. They made sure that it was so by keeping a sharp eye on their investment.

It was obvious to Josef that he could not turn to them for help with problems of character.

Nearer home, so to speak, the idea of enlisting Papa Smith, whose uncomprehending mind was not even aware of the forces he had unwittingly aroused, was amusing, to say the least. That ingenuous young man Christopher had no understanding of Isabella, and was no match for her in mental agility. The Doctor, with his sardonic outlook, was concerned only with the welfare of the body. Matthild', who, Josef considered should have been able to offer some advice on how to deal with the child — though why, he could not have said — had failed him also. He sighed as he recalled her words on the subject:

'The child has a perfect voice, Josef. She would have a perfect voice even if you had never heard of her. Oh, yes, you have trained her; you have disciplined her musically. She will go a long way. But it will be by her own intention and by her own will. Be sure, her success will have little to do with you, or you with it . . . .'

No more was said about the stupidity, or the generosity of the citizens of Strelsau. Under the Doctor's jealous supervision, and with his grudging permission, Isabella began to sing again. Quite spontaneously she sang for the joy of singing, happy in the fact that after such an illness she still could sing.

The days began to lengthen; the sun shone warmly and the snow melted on the roofs and in the streets. For an hour during the warmer afternoons the cage of singing birds was suspended from the hook outside the window of the salon, and the birds tried their spring voices against the fresh air. But they were quite silent when Isabella sang.

Within, the house was cleaned and polished from top to bottom; it could have been put under a glass dome of the type so admired by the age, and preserved for ever. Not a speck of dust dared to settle anywhere but Matthild' was upon it with cloth and beeswax. Sachets of sweet-smelling herbs were discreetly placed to counter the lingering odour of Josef's winter consumption of tobacco. In the salon the forte-piano was polished with such loving care that Isabella could see her reflection plainly in its rejuvenated flanks.

In these surroundings she blossomed, and seemed to be happy again. The bitterness so exposed in the latest display of caprice seemed to have been brought under control; she was docile and amenable. She listened with attention to Josef's instructions, and carried them out to the letter. No great effort was required. She sang perfectly. When he requested her, in the most persuasive voice he could command, to engage in runs, trills and turns to restore the full flexibility of the voice, she made no comment. When at last he brought up tentatively the subject of the projected *matinée* her new-found simplicity remained sweet and quiescent. She even seemed eager to demonstrate her art before the patronising grimaces of the Strelsau 'Friends of Music'.

The composition of the programme caused Josef some heart-searching distress. Jealous for Isabella's talents, and concerned for his own reputation as a singing master, he began by including too much. The list of items was cut down to reasonable proportions only unwillingly and with great regret.

'We must have the perfect selection,' he told her.

Isabella smiled and agreed.

'First, I think, a little Mozart, which perhaps they do not know so well. We must lead them gently. They are expecting operatic airs, so Master Mozart will help them to concentrate at the beginning. *Il Re Pastore*, I think, and Christopher shall play the violin *obbligato*. What do you think, Isabella?'

'As you say, Maestro.'

He gave her a searching glance.

'It is essential that you should be happy with what you sing.'

'Mozart will be perfect, Maestro,' she said. 'Whatever you choose, that I will sing.'

He could scarcely believe his ears, so agreeable was her demeanour. His heart was joyful as he ruffled among his books.

'And here,' he continued, clearing his throat, 'here we have some pages of Maestro Bellini, no less. *O quante volte*, the *Capulets and the Montagues*, you know . . . and here,' he went on hastily, fearful lest she might object in her sudden way to his suggestions, 'we have the splendid *finale, Bel raggio lusinghier*, from the *Semiramide* of the Master himself. Oh, how you shall sing, Isabella. Never will they have heard anything to equal it.'

He checked himself abruptly. Isabella had said nothing, but across her face there had passed a faint shadow of distaste.

He seated himself at the forte-piano. Isabella took her position before him, and the rehearsal began.

Letters of invitation, inscribed in Matthild's beautiful German calligraphy were despatched to various quarters of Strelsau, and those who received such an invitation replied in acceptance without delay since this was to be an event so rare as to be almost without precedent. Such an invitation could not be ignored. How otherwise would it be possible to judge the value of Herr Josef Weingartner's boasting?

The High, and the Not-so-High (of whom there was a small number) came in their carriages to Herr Weingartner's door to hear and see again this prodigy on whose account the Fräulein Matthild' had so completely disrupted her household when the whole of Strelsau knew for a certainty that she had little regard for anyone except her brother. It was to be hoped that her confidence was not misplaced.

The 'Friends of Music' did not compose an overlarge body; they were very select indeed. But the milling and the jostling of their carriages about the square resulted in a disruption of the normal traffic almost the equal of that when the Opera was in season. And in addition, on this occasion, there were other citizens of Strelsau, those who were not invited, but who nevertheless already spoke of Isabella with love and sympathy, claiming her for their own. *Our Isabella*, they called her. *Our little song bird*. These stood in great silent groups, eyeing each carriage and its human burden as it approached Herr Weingartner's door. They were agog with curiosity. They peered and pried audaciously. They exchanged coarse remarks with each other concerning what they saw and who they saw. But they were in the end waiting for Isabella to sing, waiting for the voice that would come to them from beyond the confines of the room they could not enter.

The 'Friends' were admitted by Maria, who bobbed and curtseyed in true country style, and received by Josef, who wore his most unctuous expression and employed his most obsequious manner. Their lackeys, descending in velvet britches and buckled shoes from their stations as delicately as their mistresses from their carriages, carried bouquets of Autumn flowers, which they presented to Fräulein Weingartner.

One would have thought, the Doctor reflects, his eyes not leaving for a second the tragic figure up there on the stage, with its dishevelled burning hair and its cold green eyes, the one word 'No' sufficing to reduce an excited audience to silence, that the young Isabella had already been a *Prima* in her own right, such had been their eagerness to see and hear her.

And how wonderfully she had sung.

The Doctor turns briefly to Matthild', whose dour features betray no sign that she is listening to a performance so extraordinary that it has brought the rest of the audience to a frenzy.

'Do you remember, Matthild', how wonderfully she sang, that afternoon? Do you remember how pleased Josef was that his efforts were completely rewarded and vindicated?'

'I told him it would be so,' Matthild' replies with that persistent reiteration so common in the aged. 'I told him she would make something of herself.'

'But surely Josef was of the utmost help,' the Doctor remonstrates gently, somewhat baffled by his wife's turn of mind.

'That is so,' she replies, 'but he could have done no more, as you well know. He did not know how to handle the child, or her talent.'

The Doctor concedes that this was so.

'I told him he was selfish,' Matthild' murmured. 'Do you not think that Josef was selfish?'

'Poor Josef,' the Doctor thinks, as his memory recreates that winter evening and the picture of the fretful, apprehensive singing master pacing to and fro before the door left so carelessly open. The memory is sharp, even after twenty years.

My pupil Isabella Weber is ill . . . . She is my best pupil. Never has anyone sung as she sings . . . .

How true that turned out to be, the Doctor reflects. At the same time he recalls his own thoughts on the subject. *How could anyone subject a child to such a stringent routine?*

Poor Josef. Poor child.

The Doctor is not aware that Madame Marchesi was moved to repeat long after him, the same sentiment: *Poor child* . . . . Although her reason for expressing her feelings in the phrase was quite different from his.

The Doctor is not able to decide whether it is because of Josef or in spite of him that Isabella has 'made something of herself'. The voice he is hearing this evening differs from that he heard those many years ago only in that there is now a maturity, a sophistication which, by the very law of Nature could not have been present then; a voice ripened by experience in the most arduous, the most relentless of the arts. The very important bemedalled and bejewelled persons from far away countries, and the citizens of Strelsau — at least, those of them who have been invited for this auspicious evening and who are now sitting in their ordered ranks below him — have been privileged to witness a unique demonstration of the twin arts of drama and music presented by no less a person than a waif picked up in the city's gutters.

By the end of the first act their allegiance is not in doubt. Their collective hearts are at her feet, notwithstanding the rumours of her arrogance, her pride, and the contempt she shows for the more unseemly of her admirers. She is, in truth, 'Our Isabella'.

The Doctor sits forward uneasily in his gilt and plush chair, ready to move at the first sign of anything untoward. His anxiety concerning her well being is by no means alleviated, and he is deeply disturbed by her own, no doubt innocent, but nevertheless ominous words: *Tonight I shall have sung my last aria . . . .* Fact? Or precognition? He shivers, suddenly cold in the oppressive heat engendered by the sheer number of the audience. Matthild', in a rare moment of understanding, reaches for his hand. She holds it gently until, with a brusque movement, he takes out his monstrous watch. Its pallid face indicates that there is but a short time to go, in terms of opera. It is a long time in terms of human unhappiness.

The *finale*, he remembers, from long ago, is unusually taxing.

And again, his memories are disturbed. What will they say, who were not born when Isabella sang? *Maestro, tell us of Isabella Weber. Who and what was she . . .?*

*

So they came, the dubious and the sceptics, to hear his 'nightingale', those whom Josef had decided could and would be the most generous in their dispensation of grace and favour. For them, as they were received into his salon, wine gurgled deliciously from the purple decanter into glasses unsullied by any blemish that might offend delicate lips. Confits were offered, displayed in little bowls, and crunched between discoloured teeth. And as they savoured these delicacies, and sipped their wine, they minced and murmured about the salon.

Towards each of them Josef displayed his personal charm; he could, it appeared, be all things to all men — and their women. He strolled benevolently from group to group, kissing the hand of such ladies as had daughters who might soon require singing lessons, complimenting them on their own continuing youth and beauty. The ladies so addressed blushed beneath their powder, and giggled.

'So droll, dear Herr Weingartner, is he not?' they said to each other.

To the gentlemen Josef bowed. To certain of them he whispered the 'man-about-town' gossip they were avid to hear, behind a discreet hand, in the certain knowledge that they could be relied upon to repeat it later, with embellishments, to their wives. Thus his — Josef's — reputation as a very knowing gentleman indeed would be enhanced. He urged further glasses of his carefully chosen wine into their by-no-means reluctant hands, in full consideration of the fact that wine judiciously applied increases the temperature of expectation, diminishes the critical faculty, and loosens the tongue. Josef was far too experienced a host to allow pools of silence to form in the stream of sociability. Silence, he would say, begets thought, and thought can lead to indecision in the minds of those who do not normally think. When, therefore, his guests fell silent for any reason Josef talked, sometimes outrageously. So mellow was his wine and so sharp his tongue (although his eyes smiled the while) that they could not long remain silent.

The matter of their conversation was to him of little importance. It was sufficient that they should talk themselves into a properly receptive frame of mind for his purpose — that purpose being the recognition of the talents of his 'backyard hen', and of his own extreme competence as a singing master.

As they talked, and chewed their confits, their eyes sought always the figure of Isabella, sitting remotely apart with Matthild'. Matthild' responded as graciously as she was able to such of the guests who condescended to speak to her, but Isabella acknowledged no one; she sat silent and withdrawn, so completely withdrawn that the room might have been empty as far as she was concerned. When the outstandingly musically pretentious of Strelsau approached her, with the intention of passing some remark which would indicate the extent of their interest in her and her talent for singing, they were disconcerted by the tilt of the aristocratic face and the green eyes that looked into their minds and stripped them of their protective layer of snobbery, and they withdrew, their remarks unvoiced. If at first they found the idea of Isabella as a singer doubtful, not to say ludicrous, her impact as a person was frightening.

The Doctor, understanding a little of the personality behind those same green eyes, enjoyed himself hugely, lifting ironic brows at each exchange of glances, savouring the shocked snatches of non-comprehension as they retreated from her presence declaring that they could by no means understand why she thought herself so different from other people.

'A quaint child,' was the most appreciative remark that many of them felt themselves able to make.

The Doctor silently applauded Isabella's self-possession and the way she ignored the conceit and discourtesy of these patrons of music. They were certainly not the first, and so far as he could predict, would not be the last to be mystified and affronted by her attitude to such as they.

To tell the truth the Doctor himself was somewhat sickened by the conceit of these self-appointed arbiters of musical endeavour. When from time to time any one of them unbent so far as to address him, fully cognizant of his post as Surgeon Officer attached to the Opera, he replied with true Ruritanian politeness, but shortly, making no effort to prolong the exchange. His mind was fully occupied with the care of Isabella. For the time being nothing else was of importance. That this particular reception was being held against his specific advice was causing him great uneasiness; so much so that he had no inclination to banter frivolities with those who were the prime cause of it, whose importance was sustained by the depth of their purses.

Josef was well aware of the Doctor's opinion, but made no reference to it. He devoted his attention strictly to his guests; he chatted; he exchanged tidbits of gossip; he agreed that the weather was still cold for the time of year, and sighed with them for the warm beauty of summer, and as he passed among them he calculated with some shrewdness the depths of those same purses and the degree of patronage they might support, to the last *Pfennig*.

The more than usually bland expression upon his large features disguised his repugnance, and indicated to both Matthild' and the Doctor that he was well satisfied with his calculations.

The flow of wine and the supply of confits being rather diminished, Josef struck a chord on the forte-piano by way of securing attention. At this signal the ladies of the select audience made their way with determination to the chairs previously set out by Maria, and arranged their voluminous skirts circumspectly about their ankles. The gentlemen took up nobly uxorious poses behind their respective wives.

Josef beamed upon them all.

'Ladies and gentlemen,' he said. 'My friends — if I may be so bold.'

The ladies fluttered their fans. The gentlemen put on solemn faces.

'My friends,' Josef continued, and they seemed not to perceive the inflection of irony that informed his tone, 'my sister and I are happy to welcome you in our house . . . .'

He paused briefly, for effect, and beamed upon them again.

'Ladies and gentlemen, I present the Herr Christopher Smith.'

They looked at one another. 'Herr Christopher Smith?'

'He is from England,' Josef took pains to impress upon them.

'Ah, England. *Aus England, ja*,' they nodded to each other, as if that explained everything. But it was clear that a number of them were quite vague as to the exact meaning of the term 'England' and its association with this particular reception.

All eyes were upon Christopher as he entered the room carrying his violin. Such a handsome young *Englisch*, the ladies thought. Such beautiful blonde hair . . . . In silence he adjusted his music on its stand. Too late he realised his *gaucherie* in not acknowledging the company, and his confusion was such that he did not realise that Isabella had neglected to observe this prime courtesy.

Josef continued as though nothing untoward had happened. But he was apprehensive.

'Ladies and gentlemen, I present the Fräulein Weber.'

A murmur passed about the company. The gentlemen leaned forward to get a closer view, if that were possible, and the features of the ladies hardened perceptibly. The Doctor rose and extended his hand to her. She turned her face to him, and her mouth softened with the suggestion of a smile. Hand in hand they went forward the few steps to her place between the forte-piano and the violin, where he bowed, and left her. The faces of the ladies became even harder.

Isabella stood straight and motionless in her green and yellow panelled gown. Her head was high. The bronze cap of her hair had been taken back from the brow and dressed by Maria in ringlets which fell charmingly on either side of her face. She looked at no one; her eyes were focused upon the picture of 'The Music Lesson' upon the opposite wall, a familiar companion for almost three years.

The expectant silence was broken only by the portentous tick of the great 'Case clock' which stood in the corner of the room.

Josef again addressed the company.

'Ladies and gentlemen, the Fräulein Isabella will begin her recital with the aria *Il Re Pastore*, of Herr Wolfgang Gottlieb Mozart.'

He sat again at the forte-piano.

*

When the audience is shamefacedly silent again after its intemperate but understandable applause, and after a pause which Andelli, with all his sense of drama, did not write into his score, the music begins again. He nurses it gently along, with many nods and glances at his musicians, who are confused by the situation, but sensitive to the importance of the occasion and their part in it, and certainly do not require all those nods and glances. Nevertheless, both Andelli and the orchestra are preparing themselves for the dramatic scenes which lie immediately ahead. Isabella moves a little way upstage the better to confront the villainous Duke; in the strong light her eyes seem to be softer, but her expression is still quite disdainful. The Chorus intones very softly the beautiful variation on the theme *Non mi tentar*, cunningly interwoven with *Te amavi — I have loved you*. This repetition is intended to convey the idea that she might relent, should he listen to her with understanding.

In exquisitely modulate phrase, so pianissimo that the attention is strained to catch it, she repeats the ultimate line of the aria: *You I once loved* . . . .

The tone is so precise, so pointed, that it does not fail to reach all the corners of the House, even to the Royal Box.

The precise second on which she ceases to sing, however, is not discernible by any person in the audience, for on a sudden *forte* from the orchestra the Duke takes up the air, approaching her as he does so.

. . . *On that golden day*, he sings, going on to declare his great joy at that time; how he too loved, and still loves her; how happy they were for a short time, until events which could not be gainsaid overtook him and he was compelled on the hour to leave his Dukedom, etc., etc. He comes closer, but Cordelia backs away from him, her arms outstretched to ward him off.

Cordelia, he pleads, *I love you still*. But his mien is far from pleading.

Cordelia draws herself up to her full height; her eyes glisten in the spotlight. She looks upon him with scorn, in disbelief.

False. False. You have never loved me. Your words and your heart are false.

So bitter is the tone, so hard and clear the enunciation, delivered at arm's length, that the luckless Duke feels his own professional force weakening, as if he had been in truth guilty of the supposed crime. Even if he were not mumming, Giovanni Tortoni playing the Duke is quite sure that he is no match for the likes of Isabella.

For his Majesty, sitting up there in the Royal Box, the situation is much too realistic. What an actress she has become, he thinks. Yet he shifts uneasily in his chair and glances round at his gentlemen *aides* in the hope of catching some unforgivable expression on one face or another.

Amelia Viktoria perspires, and agitates her fan before her almost crimson face.

On stage, Cordelia stands proudly before her betrayer.

How beautiful you are, Cordelia, he coaxes her.

But she tosses her head in contempt at his attempted blandishment.

'How proud she is,' the audience thinks to itself, but it remains silent, having experienced a small taste of Madame Weber's forceful personality — those glittering lords and perfumed ladies, and the lesser breed that fawn upon them. That one little word, and that basilisk stare from those green eyes had brought home to them forcefully the coarseness of their behaviour. Yet all they had wished to do was to express their appreciation of her singing, and what better way was there than the one that was customary, as far as they knew, in all opera houses. Certainly it was so in their own. They had hardly believed the rumours that were circulated of her arrogance, her disdainfulness, her insolence towards such as considered themselves her social superiors. It could only be malicious gossip, they said to themselves, the result of jealousy, or at best, envy. They were quite sure that a person so dependent upon public favour for her living could not, would not, behave so.

*

That was the way of it in Herr Josef Weingartner's salon on that far away April afternoon. Face to face with a child about whom they thought: 'Perhaps she can sing, after all. It is to be hoped that she can sing, for there is little else to recommend her,' they had remained completely silent, more intimidated by her unsmiling face than they would have cared to admit. But at the end they knew she could sing, although they could not have said how, or in what manner she sang.

When it was over, the applause that had begun desultorily died away abruptly as Isabella, ignoring what she considered to be a useless demonstration, curtseyed only to Herr Josef, and returned to her place by Matthild' and the Doctor.

Josef was at once relieved and disturbed. Isabella had certainly acquitted herself well, as he had expected, and she had behaved in the best possible way. But there had been all the time a gnawing fear in his mind that all was not as it should be. She had been too calm, too self-possessed; she had sung so perfectly that his heart had sung in sympathy. He had not been able to detect a single flaw throughout the length of her recital, but it had been the recital of a machine called Isabella Weber. Like a statue she had stood, neither seeing nor acknowledging the presence of the company. Not even to all-too-obviously spontaneous acclaim which had emanated from the grey ghosts in the street beyond the window had moved her although one or two, bolder spirits than the rest, had called her name in the frenzy of their worship, causing frowns expressive of the sharpest distaste to pass over the features of the more select company within.

'Isabella.' They had called from beyond the window. 'Isabella . . . Bella . . . Bella . . . .'

Their voices had fallen to a murmur as Matthild' appeared at the window. This time, however, they did not go away. They were waiting, as they considered it their right, to see the great ones leave the house and enter their carriages; to hear their cunning *schön, aber schön's*, their interminable *wiedersehen's* and their meaningless *Grüss Gott's* . . . . To observe and comment upon the Herr Josef's endless bowing and scraping, for all the world as if he had been one of the newly-invented automatons now amusing the public.

When finally the guests had departed, unable to find reasons for still lingering, and the House of Weingartner was recovering something of its normal appearance, Josef was moved to embrace his pupil.

'Thank you, Isabella,' he said, with such simplicity and such feeling that for the first time in many days she smiled upon him naturally and happily.

'Thank you, Maestro,' she replied in her turn.

'There is little doubt,' he assured her, 'that your immediate future will be well taken care of. How far you will go thereafter will depend upon yourself.'

'I am to go to Vienna, then? To Madame Marchesi?'

'To Vienna indeed, to do what you will, for better or for worse, Isabella.'

The Doctor, his expression quite unhappy, looked at Josef sharply, but Josef's own features were quite benign.

When Isabella enfolded Papa Smith, who had sat silent in his corner throughout the whole afternoon, in an embrace more loving than any he had received before, there were tears in the eyes of both of them.

'Dear Papa,' she whispered, 'I love you. Thank you for everything.'

By way of further preparing the ground for Isabella's descent upon the Capital, Josef re-phrased and re-wrote the letter which she was to carry with her to Madame Marchesi. In spite of Isabella's triumph, he wrote in a heaviness of spirit, and with a sense of loss.

'Dear Madame Marchesi,' he began, with a fine flourish of the pen, 'It is my wish that my pupil Isabella Weber should give this letter into your hands in person. Thus you may be able to form an immediate impression of her. First impressions can be so important. As for her musical ability, I myself have nothing but praise. I consider her technique on the forte-piano to be of the highest quality. Her grasp of the theory of music is beyond question. As for her voice, no words of praise could be considered excessive or exaggerated. She has a brilliant voice, lyric in quality, but with a sonority and a volume (when needed) of the true dramatic soprano. It is a voice quite outside my experience. It is so perfect that the points of junction of the various registers are impossible to detect; neither can a conscious thought be apprehended in the placing of notes. I shall not tire you, dear Madame, with details of her *tessitura*; these you will discover for yourself. The basis of her technique is her phenomenal breath control, which in turn is the secret of her incomparable *pianissimo*, which itself is scarcely more than a sigh, but which can penetrate to every corner.

'She sings, in sum, with simplicity and naturalness . . .'

Further to this preamble he added, with a rare humility:

'. . . I need not say, dear Madame, that these phenomena have little to do with me. Something I have been able to teach her; the rest is of God . . . .'

From this point he continued with increased misgiving, conscious that his understanding of the human soul was, as had been so often pointed out by both Matthild' and the Doctor, short of completeness.

'I fear that as a person she may present you with problems. Certainly I have to confess that in dealing with this aspect of her development I have failed most miserably. Her pride and her stubbornness (perhaps the same thing), together with the humility with which she appears to accept instruction only to question it too pointedly thereafter, as if to show the futility of the instruction in so far as it concerns her, present so complicated a character that a mere man cannot begin to understand it. The irrationality of behaviour in one who is so intelligent surpasses by much what I conceive to be the notorious "artistic temperament".

'On the other hand, it may be that during the natural process of "growing up" she needs the sympathy and intuition of another woman, an artiste such as yourself, dear Madame Marchesi . . . .'

Madame Marchesi smiled her little smile at the thought of what it must have cost Herr Josef to unburden his heart in such a manner. She was well enough acquainted with him to know that whatever the circumstances he could never forego his outrageous flatteries. That apart, however, it was not long before she understood, to her own dismay, that where Isabella was concerned he was a serious and deeply disturbed man, and that his dilemma was far from being imaginary.

She decided that she could make no reply until she had seen and spoken with Isabella herself.

*

Looking down from the window of her apartment on the Kärntner Strasse Madame Marchesi watched the *fiaker* draw up at the portal with such a portentous flourish and jingle of harness, such a stamping of hooves and such a prodigious snorting, that it might well have signalled the advent of some minor sprig of royalty. The coachman, voluminously wrapped even though it was well into the month of May and quite warm, descended with a great effort from his box and began to labour at a small mountain of baggage, goaded to his task by a volley of commands and instructions from a plump, rosy-cheeked servant girl who herself had almost fallen from the carriage in her haste. A young man — tall and fair in colouring, Marchesi noted with interest, and handsome in a weak kind of way — had also descended, and was offering his hand to the third and last occupant of the carriage, to whom Madame now devoted the sharpest attention.

This person was attired in what Marchesi could only describe, with something of a shudder, as the 'country fashion'. It was quite out of place in modish Vienna. Madame Marchesi sighed at the thought of what lay before her in order to transform this duckling into a swan. Yet the girl stood tall and straight, in spite of her dowdiness, beside her escort, and surveyed the activities going on about her as though she were fully accustomed to that kind of attention. The thin pale face and the rather prominent nose certainly bore no stamp of peasant stock. Pride there was, a pride that was the manifestation of an intrinsic natural authority. Josef had indeed not exaggerated his misgivings.

Beauty there was not, as yet. But time would arrange that little matter too.

The activity about the *fiaker* continued unabated until the baggage, the transference of which had been somewhat impeded by curious passers-by, had been piled neatly within the portico by the coachman, who was now perspiring freely in the spring sunshine. He was still being urged to greater exertion by the servant girl, who was requesting him in a shrill country dialect to now carry the heavier parcels up the stairs, a thing he was adamant in refusing to do. Madame Marchesi was fascinated by the scene. The servant was clearly, in her peasant way, the mirror of her young mistress, she thought. She did not know that admiration and worship on the part of the servant, and complete trust on the part of the mistress, had bonded the two closer than sisters. Nor was she aware that the servant had abandoned her home, her parents, and her privileged position in the house of Josef Weingartner to follow the doubtful fortunes of a would-be opera singer.

The first, but happily not the most consequential, mistake that Mathilde Marchesi made in her disciplinarian way was to treat Maria as a servant.

She made that mistake only once.

The business down below having been more or less satisfactorily settled, the little party was received by Madame Marchesi in her salon with a warmth that was typical of her in spite of the fact that many of her friends, and all of her pupils considered her to be a relentless martinet; a dragon, they said, without feeling.

Face to face with this dragon in the somewhat depressing Viennese *décor*, Christopher introduced himself and his companion.

'Christopher Smith,' he said, making a bow. 'May I present the Fräulein Isabella Weber?'

He remembers even now the look of surprise on Madame Marchesi's face. Clearly Herr Weingartner had omitted to advise her of this situation. He hastened to explain in his halting German, feeling rather foolish.

'The Fräulein is my sister, by adoption, Madame. She was adopted by my father when she was but a mere child. We are in all things, in effect, brother and sister.'

Madame Marchesi seated herself with deliberation in the ugliest but most comfortable chair in the room. It was perhaps coincidence that the afternoon sun shone obliquely upon Isabella.

'So, Herr Christopher Smith,' (she pronounced it *Schmeet*), 'what then are your intentions?'

'I am to study the violin at the Conservatorium,' he told her, feeling relieved that he at least would not have to face those intense eyes every day.

'Indeed,' she said. She looked them back and forth as if revolving a problem in her mind. When at last she spoke, her words appeared to Christopher to be something in the nature of a Sibylline utterance.

'Perhaps . . .' she said. 'Possibly . . . but unlikely . . . .'

Like the Sibyl she stopped short of explaining what it was that was unlikely.

'The Fräulein,' Christopher continued, his courage falling rapidly, 'wishes to study singing, with you.'

'So much I have been informed by Herr Josef Weingartner,' Madame Marchesi told him dryly. 'It is arranged. Before we come to either playing the violin or singing, there is much to discuss. Not least, there is the question of your accommodation. I had not understood that you would be to all intents and purposes a family. It is fortunate that the apartment on the floor above is vacant, and at my disposal. It has sufficient accommodation for you and your servant. The expenses, I am given to understand, will not overly concern you. I presume you will not object to live so near to me?'

It was a statement rather than a question.

'I shall be happy to be near you, Madame.'

Isabella spoke for the first time since their arrival. Her voice was gentle, her enunciation precise. It betrayed no hint of vocal splendour. Madame Marchesi remarked the phrase '*I* shall be happy,' and smiled upon her.

'The apartment then is yours, from this moment. It is apt to be something noisy at times. The Kärntner Strasse is a main thoroughfare.'

'That is of little importance,' Isabella said, thinking of the traffic about the square at Strelsau, and the constant booming of the bell in the tower of the Cathedral of St. Michael.

'So. That is arranged. Now we will attend to the disposal of your baggage, and then we will take tea, in honour of the English Herr Christopher.'

At first sight even Madame Marchesi had been in some doubt, like the 'Friends of Music' at Strelsau, about Isabella's potentialities as a singer, especially as a singer of opera. Like them, she could find little promise in the thin body, the pale face crowned with a mop of red hair, and the prominent nose. Furthermore, Marchesi had to admit that she could not understand the child. The second part of Josef's letter of introduction bore the stamp of truth in its reference to the problems that faced her concerning Isabella's personality. At all times the child was a law unto herself. For the first three months of their association she had not made a single friend. She neither smiled, nor engaged in conversations; she remained coldly aloof from the frivolous female chatter that surrounded her day by day.

Madame Marchesi treated her gently at first, and pursued her normal training course just as with her other pupils, and there was no occasion when Isabella showed any wish to do other than please. She obeyed every instruction; she made no criticism about that instruction, as had happened so many times during Herr Weingartner's lessons. Her scales, arpeggios, her turns and *appoggiaturas* were as clear as a bell; whatever she was given to sing, she sang, so naturally, so spontaneously, and so perfectly that Marchesi was in despair. It was almost beyond belief, and but for the fact that she, the foremost singing teacher of her time, was hearing it with her own ears, she would not have believed it. Herr Weingartner, so far from exaggerating, had not told the half of it. There was nothing more to teach Isabella about the technique of singing. All that remained to do was to ensure that she had the right social and musical contacts.

At the end of the first three months, Madame Marchesi wrote in her turn to her friend Josef. She wrote guardedly, but confessed that she had no solution to the problem of Isabella. The writing of the letter caused her as much agony as Josef had experienced in the writing of his.

'. . . I perceive,' she wrote, 'that the Fräulein Isabella Weber is an original, at once unbelievably adult in so far as her singing is concerned, but with the qualities of a charming child. Your reference to the effects of ancestry and inheritance — which I find of the greatest interest — has, I am sure, some foundation of truth. It would certainly be much more helpful if we knew something of her forebears.'

(Josef shook his head dismally at this. Alas, a slum child taken in by a quixotic Englishman . . . who could possibly say what blood ran in her veins?)

'For the rest, it is possible that she is of the age at which all the uncertainties of life are compounded. This, and the lack of companions of her own age, whether of her own social standing or not, has probably contributed largely to her own uncertainty regarding social balance and understanding.

'It is to be hoped that such a state of affairs may be quickly rectified. With me she will associate with young people of similar age and aspirations, although I regret to have to say that their musical knowledge *and* their ambition linger somewhat behind that of Fräulein Weber. She will be introduced to a wider circle of people with musical knowledge and ability than was possible, I believe, at Strelsau. Thus she may develop, in *addresse* and sophistication.

'There is, attached to what I call jokingly 'my little company', a young Italian gentleman by the name of Ruffino Andelli, who will one day make a name for himself as a composer. At the moment he acts as my accompanist and *repetiteur*, in which capacity he also learns much about the difficulty of writing music for future ''*prime donne*''. He seems greatly attracted by the Fräulein Isabella, preferring to accompany her rather than any other of my pupils.'

(Madame Marchesi's countenance was as serene and innocent as could be as she wrote these words but, when Josef read them later in Strelsau, his countenance was dark and grim, as she had anticipated.)

'Signor Ruffino,' she continued, 'has already declared quite openly that he will write an opera for her. I myself have no doubt that it will be so.

I agree with your assessment of the young lady's talent, dear Herr Weingartner. About this I shall have more to report at a later date . . . .'

And that was that, Josef thought. The sound of his sigh was loud in the empty salon.

The young ladies who comprised Madame Marchesi's 'little company' considered Signor Andelli to be egotistical and theatrical. He was Italian, and darkly handsome in the Italian way. He was so *gallant* that they could not forebear to tease him mercilessly, and they laughed at his pretentious ambitions. Only Isabella did not tease or laugh. She encouraged him as he struggled with his composition, although she fought with him to the end of almost every phrase. When at last, singing her part from his notes, she declared that such and such a passage, or indeed such and such a note, was unsingable, he sulked, but rewrote it according to her direction. The opera *Alma Perdita* was as much Isabella's as Andelli's.

To Isabella he dedicated the opera. So much he loved her, so much was his genius interwoven with hers, that his music was so faithful a portrait of her that they might have been one and the same person.

When, in so short a time, his idol stepped off her pedestal (in his opinion) he was so sick at heart that he never wrote another note for her, and even destroyed immediately the first pages of a second projected opera. 'Why should I play the clown?' he asked himself in his torment. 'Why should I laugh and make music with a broken heart?' But he was a Latin and quite resilient. He wrote other operas for other ladies who, if they could not sing like Isabella, at least adored his Italianate person. These works never approached the perfection of *Alma Perdita* and they were fated to decline after the first presentation. Thereafter he composed only purely orchestral music. But he still loved Isabella.

Alma Perdita was written and rehearsed by the students of the Conservatorium, and was mounted at the end of the student year, sung by Isabella and conducted by Andelli. It was his first public appearance, and her first public performance. Neither of them could have imagined what the result of this juvenile *tour de force* was to be.

Internal concerts at the Conservatorium were of little importance, except to those taking part. For the big yearly production, however, friends and relatives were always invited, as well as men of influence in the musical world. As a consequence of this favour Herr Josef Weingartner (Herr Smith being unable to leave his desk at the opera at Strelsau) braved the winter journey to Vienna, and found himself sitting in the same audience as Herr Hans Richter, although at that time the two men were not personally acquainted, a situation shortly to be rectified. Matthild' had stayed behind in Strelsau, unwilling to travel so far in such inclement weather. She was enjoying in her own way the devoted attention of the Doctor.

For those engaged in whatever capacity in the production, the pressure of work was an increasing nightmare. Scenery was to be built; 'props' were to be got together; costumes were to be made; the orchestra was to be rehearsed whenever a sufficient number of musicians could be assembled in one place. When this could be arranged, the way was far from smooth. So unused to such music were the student fiddlers, so outrageously hypercritical was Isabella, that Andelli despaired of ever getting his opera presented. He remembered even now, as she sings up there on stage, how he approached the evening with fear, in spite of his show of Italian bravura. He remembers in particular, his astonishment at Isabella's self-possession, her composure as she walked into the scene as the curtains rose at the end of his overture and quite simply sang, as if it were an everyday business with her, as if there had never been those endless heated discussions and quarrels, those acrimonious arguments about phrasing or tempo or *piano* or *forte* . . . . He remembers vividly the last heated exchange they had had, at the final rehearsal after he had spent a wakeful night at his forte-piano, revising his ideas of Isabella's voice. It was in fact this very aria, concluding the second act. He had with the best of intentions stopped his orchestra and made some remark to her concerning the phrasing of a certain passage in the course of the aria. He was quite unprepared for what followed.

For a full minute the stage was frozen.

Isabella looked down upon the hapless Andelli. Her eyes were green as they reflected the footlights.

'Yes, Signor Andelli?' she enquired with something like resignation in her voice.

Signor Andelli swallowed. He pushed a hand backward through his flying hair.

'Fräulein . . . .' he said. 'Fräulein . . . .'

'Yes, Signor Andelli?' She was all sweetness. 'Am I then singing something wrongly?'

'Indeed not, Fräulein. You are singing beautifully, Fräulein.'

'Well then, Signor, why have you interrupted me?'

Now both her face and her voice had changed.

'The aria, Fräulein.'

'What of it?'

'We are near the end of the aria, Fräulein. You have been ill-used; you have been seduced and abandoned. But your lover has caught up with you; you are angry, and afraid, yet you sing as if it were a lullaby. You must give me anger, and fear, Fräulein.'

Isabella remained coldly silent for quite a long time. When at last she spoke it was Andelli who felt the fear.

'Should you not wish me to sing, Signor Andelli, I must remind you that I do not have to. There is an understudy, I believe.'

'Fräulein . . . Isabella. What are you saying?'

'I know this work by heart, Signor Andelli. I have lived with it for many days and nights. I sing it as I feel it. *I* am Cordelia, from the beginning to the end of the opera.'

'I too know the work by heart, Fräulein. I am the composer, should you have forgotten.'

He was quite sarcastic.

'It is agreed,' Isabella replied. 'You are indeed the composer. But I am the singer. I have to sing your music.'

She was equally sarcastic, but still in full possession of herself. Nevertheless they were behaving like children.

Andelli fumed, and threw his arms about. He became at once the outraged man and the outraged composer.

'You will sing my music as I require it to be sung, Fräulein.'

Isabella was unmoved by this declaration. She continued to study him coldly.

'You may have written the music, Signor, but it is abundantly clear that you have very little idea of the task involved in the singing of it. And while we are about it, I have something to say that may or may not be helpful to you. I shall sing your music the better, and the words that go with it, when you have learnt to conduct it. Your orchestra has already made several unfortunate changes in tempo which perhaps you have not noticed. This is because they are listening to me instead of paying attention to the waving of your little stick.'

Andelli's dark face was pale with frustration.

'*Madre di dio*,' he burst out. 'Who is it that shall be important here? I devise the opera. I compose the music. I, Andelli, conduct the orchestra, while you . . . you . . . .'

Isabella interrupted him, very gently.

'I sing the music. Without me you would have no opera worthy of the name. You know it, and I know it, Signor Andelli. You must certainly listen to my singing, if you are to make anything of your opera, but while listening you must endeavour to convey to your musicians your understanding of what I sing and how I sing it. By means of your little stick,' she added.

From the height of the stage above them she smiled down upon the crestfallen musicians.

'Thank you, gentlemen,' she said. 'Your personal attention is much appreciated.'

Turning to Andelli, she requested, without the smile: 'Now, Signor, shall we go on?'

That was the last of the disagreements between them.

Andelli tapped his desk, and raised his arms sulkily.

Isabella resumed her proud stance before the villainous Duke.

'The aria, bar 1, if you please, gentlemen.'

The Duke repeated his line for the run-in, coaxingly . . . .

'*How beautiful you are, Cordelia* . . .' so that Cordelia might enter on the aria *Non mi tentar*'.

Although the performance was in a sense strictly private, gossip knows no boundaries. It was not long before the whole of Vienna was in no doubt that a new star had been discovered. In the streets, in the cafes, and in the salons, it was Fräulein Weber this and Fräulein Weber that. Even Madame Marchesi, who was not normally given to exaggeration, was stimulated sufficiently to write in her diary before retiring very late, that same evening:

'Tonight the Fräulein Weber surpassed my most sanguine expectations. There is not a voice like hers to be heard in the whole of Europe. Unfortunately her personal attitude leaves still a lot to be desired. She omitted, for example, to acknowledge in a fitting manner the rapturous recognition by the audience of her performance. This was a great shock to me, and I am afraid that the adulation she received in consequence of her performance was by no means calculated to diminish the arrogance which seems to be an integral part of her nature. Poor child . . . .'

Isabella's admirers did not give that arrogance a second thought. With a voice like that . . . .

'Ah,' they will say, telling their children, and their children's children. 'But you should have heard Isabella Weber. Now here was a singer. No one could sing as she sang, no one ever will. Hers was the last voice God made.'

They were quite sure about it.

They will chuckle at the conceit of other songstresses whose voices were like the croaking of frogs compared with the notes of the nightingale and they will shake their sage heads as if, though willing to consider the possibility of the existence of other singers, they had to admit the futility of it.

'We have heard them all, one by one,' they will say. 'We have heard your Pattis, your Sallas, your Gersters, your Bernsteins, your Reigels . . . . Not one of them could hold a candle to Isabella Weber . . . .'

They were very loyal; they knew her ability. They even knew her worth.

Madame Marchesi knew it too, the old dragon. She knew what she was about when she kept the Herr Direktor Franz von Juaner of the Imperial Opera dangling yea-and-nay for three months longer than was necessary in his efforts to get Isabella's signature on a contract. It was of course to get a fatter fee, for well as Marchesi understood such men, she understood the value of money more.

It was the same in her dealings with Herr Hans Richter, very soon after the presentation of *Alma Perdita*, and the story of this encounter went the rounds of Vienna for a long time. Herr Richter was the young up-coming new man at the Opera. He had ideas for the future of his own career as well as that of the Opera. Naturally he wanted the best; he wanted Isabella — for his new *Fidelio*.

Everyone knew about Madame's cavalier treatment of poor Hans Richter; such an inoffensive little man he was, with his execrable Viennese accent.

He came to Marchesi in a state of melancholy.

'Madame,' he pleaded, 'we are advanced in preparation. Would you cause utter confusion, and my downfall, before I have well begun?' (He had not long been at the Opera House on the Ring). 'Consider, Madame, I beg of you, my first production in this city . . . .'

'You go too fast, Herr Richter,' Madame told him. 'The Fräulein Weber is but a child, in her twentieth year only, and quite inexperienced. She knows nothing of love, nothing of marital fidelity such as Herr Beethoven wrote of.'

'Nothing of love, Madame.' He was quite dumbfounded. 'I have heard her sing, Madame. I was present, by your invitation, you may recall, when she lately sang *Alma Perdita*. And you tell me she knows nothing of love. Pfui, Madame . . . .'

Marchesi was dumbfounded in her turn. Was it possible that this comical little man had detected something in Isabella that had passed her, Marchesi, by? It would bear looking into.

Meanwhile, she drew on her reserves.

'The voice is a precious instrument, Herr Richter, and must be used with care, as you well know. Too much too soon, and — pouff. Besides, you already have my pupil Louise Kaulich, *and* my Ida Baier . . . .'

'Fräulein Kaulich has a very pretty voice,' Herr Richter conceded, 'but she is not, in my opinion, *Leonore*. The Fräulein Baier has a remarkable talent for the histrionic art, but — my profound apologies to her — and to you, my dear Madame Marchesi — she cannot sing. I must have Fräulein Weber.'

'I perceive that you know your own mind, Herr Richter.'

'Madame, the matter is of the greatest importance. We have our other principals; we have our chorus; the gentlemen of the Musikverein are rehearsed to the last note. We lack only the lady. Madame, the whole of Vienna hangs upon your decision.'

'That may well be so,' she replied. 'The expectations of Vienna matter little to me, Herr Richter. And I do not give my jewels to the first importunate who asks for them.'

Richter was so shocked by this forthright statement that he could say nothing for some time, and when Madame reminded him of the little matter of the fee to be written into the contract they were supposed to be discussing, he uttered a groan that seemed to come from the soles of his boots. He loosened his cravat with a trembling hand. His world seemed at that moment to have stopped.

'M . . . M . . . Madame . . . Madame is pleased to amuse herself,' he stuttered at last. 'Madame cannot be serious, surely.'

Madame looked down a nose as fully patrician at that moment as Isabella's.

'Madame was never more serious, Herr Richter.'

'But such a sum . . . . Never has such a sum been paid to a mere singer.'

'The Fräulein Weber is no mere singer. In any case, there is a first time for everything. That time is now upon you, Herr Richter. The more perfect the jewel, the higher the price. The Fräulein Weber is unique; she has no equal in singing or in acting. You yourself know that such persons do not present themselves everyday, and when by chance they do, you must be prepared to pay for the privilege of using their gifts, especially in forwarding your own prospects.'

Herr Richter mopped his perspiring brow; the day was very warm. But Madame watched him with cool detachment.

'After all,' she continued, 'should you feel that she is not worth such a sum, why should we continue? Why should we consider Vienna for the Fräulein's debut, a city wracked by schisms and international quarrels, a city divided by class distinction and religious prejudice; a city in which there is a distinct lack of musical scholarship and appreciation? There is time enough for many things to happen, and one day Fräulein Weber will have the world at her feet. Meanwhile, there are other opera houses. There is Prague, Dresden, Munich. There is Baden . . . .'

'Madame, I beg of you . . . .' Poor Hans Richter was near to tears.

Marchesi went on relentlessly.

'There is St. Petersburg, there is Paris. There is even London.'

'Madame, Madame, now I know that you joke. Whoever heard of music in London? That is fantasy. But such a sum . . . that too is fantasy.'

She was a clever woman, was Marchesi, and she enjoyed the game. She was not overly concerned about money, as her detractors claimed she was; quite simply she knew to the last *Pfennig* the value of her merchandise. Isabella learned a good deal more than singing from Marchesi. Those were the days before music lovers paid much attention to self-appointed critics, the days when singers sang, and audiences paid for the sheer pleasure of listening to them without too many protestations. Most singers accepted gratefully what they were offered. Under Marchesi's tutelage Isabella demanded that little more. Audiences used their own judgment, and those who knew, knew what they liked.

They liked Isabella Weber. So much was quite evident. Not only did they like her, but they were quite willing to regard her foibles as evidence of an intriguing eccentricity, and a basis for conversation. Gossip concerning the nature of this conversation reached Isabella's ears from time to time, but she paid no attention to it. Very little that was said about her caused either pleasure or distress. The important matter was singing, and in that she was always more concerned that she had satisfied herself rather than what others said about her.

In singing *Alma Perdita* for the Conservatorium Festival she had made more ripples on the surface of the pond of Fate than she was aware of. She had brought great joy to the heart of Herr Weingartner and happiness to Signor Andelli; she had aroused reactions which varied between jealousy and hatred to admiration in her fellow students, and she had fed the ambitions of Hans Richter. Only Marchesi remained outwardly unmoved. She still hoped to understand Isabella one day.

That day of understanding was brought no nearer by the impact of Isabella's first professional appearance on the stage of the New Opera House on the Ring under the baton of Hans Richter. He, poor man, was initially convinced that the furore of applause at the conclusion of the piece, and the later acclaim for the production, was for him, a recognition of his perspicacity in discovering and bringing to the operatic scene such an artiste as Fräulein Weber. He had expanded visibly at the curtain call, unaware of Marchesi's ironic grin, unaware that the reluctant but proud appearance of Isabella had provoked the audience to a behaviour frenzied and verging on the riotous, unaware even that the gentlemen of the orchestra, those blasé democratic stalwarts of the Musikverein, had risen to their feet for her and applauded intemperately. It was only at rare intervals that they heard singing of such quality.

Isabella herself showed no reaction. She stood silent and alone on that vast stage as the curtain rose for her call. So small and fragile she was, but with head held proudly high, before the grim prison walls looming prophetically about her, and the great shadowed auditorium before her where they clapped and gesticulated and called her name ceaselessly. She stood unmoving before them as the demonstration increased in volume and lengthened in time, giving no sign of acknowledgment or appreciation of their fervour. When, with perhaps a suspicion of impatience, she turned and walked off stage, the curtains closing as she went, they were still clapping and calling her name.

In that uproarious House, Herr Josef Weingartner alone groaned in spirit.

'Dear God,' he thought to himself, 'why have I condemned the child to this? What have I done?'

*

For the next forty-eight hours Isabella was inaccessible. She lay for that time in her room, exhausted. She was devotedly attended by Maria, who with salty whip of tongue and peasant obstinacy kept at bay all who inconsiderately wished to inflict themselves upon her, to present gifts and flowers, to laud her, to repeat to her how well they thought she had sung, to the point of nausea. Isabella kept her thoughts to herself. Only to Maria did she speak of her distaste for the vulgar crowds, albeit dressed in jewels and immoderate costume, who shouted and screamed their desire to possess her, to bring her down to their mindless level, to make her one of themselves. Not even her 'family', Christopher nor Herr Weingartner, were able to gain admission to her presence. Herr Hans Richter, as full of guile as an ancient Greek, came bearing gifts, and wished to tell her how well they had done. He was turned away by a more than usually belligerent Maria. He was less capable of understanding than Herr Josef. Only Marchesi made no attempt to burden Isabella with her presence. There would be time later.

*

The year was drawing to its close. The weather was more inclement, colder than usual, with high winds and drifting snow. The people of Strelsau ignored these phenomena as best they could; it was after all a regular occurrence. Besides, preparations were in hand for a great gala. At the Castle there was to be a splendid dinner, and a Ball to which all the important people of Ruritania were to be invited. There was to be music, and dancing which would last throughout the night. In between these two events there was to be a Grand Performance at the Opera. In the streets, for the delectation of the commonalty, there would be bonfires, and flags flying, and wine free from the barrel, and dancing by such of the townsfolk as could stand upright at the end of the day.

This *Fest* was to take place on the first day of January, of the year 1877. The occasion was the twenty-first anniversary of the birth of His Highness Karl Ernst Maria Michael Frederik, Hereditary Grand Duke of Ruritania.

A handsome, well-set up young man was the Grand Duke, particularly when he was in uniform, a form of display he affected more frequently than was absolutely necessary since he was quite aware of his physical assets. When he rode to the hunt accompanied by his cronies, or when he drove through the streets in a gentler form of recreation, he would smile and bow to all whom he perceived were admiring him, and the eyes of the young ladies in particular would follow his carriage wistfully until it was out of sight. Then they would go back and dream their dreams at the kitchen table or the shop counter. In Society the young ladies were besought by anxious mothers to make themselves agreeable at any function where it was possible that Fate, assisted by a sophisticated hostess, might bring about an encounter. To do the young ladies justice, they showed no reluctance in conforming to this request.

In their boudoirs, by way of preparation for this possibly all-too-brief encounter, they assiduously studied the forte-piano and singing (much to Herr Weingartner's personal profit), the gentle art of water-colouring. Sometimes they would even read a book, a selection perhaps of romantic poems, which alone would ensure an interesting if daring conversation. It was well known that the Grand Duke was an amateur of the arts as a respite from hunting. In particular, in all things respecting the Opera, he took a keen and lively interest. For his anniversary, he let it be known, there should be presented something new and exciting, since he was weary of the traditional performances. This request was conveyed to Herr Weingartner, in his capacity of arranger of such affairs, by the Directorate.

The situation was quite unbelievable. Returning to Strelsau from Vienna, he lost no time in putting his ideas before the Directorate . . . . He would arrange, he told them, a grand production of the new opera (much talked about in that city) *Alma Perdita*, with Isabella in the role of *Cordelia*. The piece should be conducted by the composer himself. At one stroke, he told himself, metaphorically rubbing his hands together, he would achieve his ambition.

He had not expected such an immediately dusty answer. A new opera. Why a new opera, they wished to know, when they had all the music and a good deal of the scenery of the old ones ready for use. They discussed ways and means acrimoniously; the ever-present question of costs was tossed to and fro with bitter words. There was at one stage outright objection and a refusal to participate in such a mad scheme. Who was this Fräulein Weber, that she should be accorded such an honour? Who was this Signor Andelli, that the scribblings of his nonage should be preferred before established and well-loved operas? The people of Strelsau would never accept it. Modern music might be all very well for Vienna, but at Strelsau . . . . Not even the Grand Duke himself would countenance such a thing . . . .

So bitterly adamant were they that the matter was indeed referred, at Herr Weingartner's insistence, to the Grand Duke himself.

It was with some patience that His Highness confronted Herr Weingartner in his office at the Opera House.

'There seems to be some conflict of opinion, Herr Director,' he greeted Josef. 'You are paid to arrange these matters, not to burden my private life with your disagreements. This opera you propose to inflict upon us, this . . . *Alma* . . . . How do you call it . . . this *Alma Perdita*. We have never heard of it.'

'A somewhat meaningless title, Your Highness,' Josef explained. 'Titles as such have in general little value. They rarely convey the true depths of a work. *Alma Perdita*, Highness, is a new opera, completely new. It was in fact only completed earlier this very year, but it has been presented with great success in Vienna. Not professionally as yet, you understand.'

'And the composer? I take it that we have never heard of him, either.'

'That may well be so, Highness. The book by Signor Giacomo Giacomini, was set by one Signor Ruffino Andelli.'

The Grand Duke frowned.

'Both Italian citizens, no doubt,' he said. 'They are legion. A disreputable body of men, I have been told, especially if they are musicians.'

'That may well be also,' Josef repeated himself. 'The Signor Andelli is at present a student at the Vienna Conservatorium.'

His Highness glared.

'Great God. Is that the best you can do, Herr Director Weingartner? On the day I attain my manhood you propose to entertain me with student scribblings?'

'Your Highness could scarcely be further from the truth, as you will realise when the piece is mounted.'

'But students, Herr Director . . . . However sincere they may be, it cannot be denied that they lack a certain . . . .'

He paused.

'You did mention the word 'students', did you not?'

Josef permitted himself an unctuous smile.

'Your Highness misapprehends me,' he declared. 'This work will be performed for you by a fully professional cast, with the exception of one who was herself lately a student at the Conservatorium. A turn in the wheel of Fortune has decreed that she is no longer a student, but a singer more professional than all.'

'You exaggerate, surely, Herr Director. How can this be?'

This . . . student, had the good fortune to be heard by Herr Hans Richter of the New Opera in Vienna. For him she sang three performances of Herr Beethoven's *Fidelio*. She was immediately acclaimed. Her name is already a household word in Vienna.'

'Indeed! The name of this paragon, if you please, Herr Director.'

'She is the Fräulein Isabella Weber, Your Highness. You may be interested to learn that she was my former pupil, and a native of Strelsau.'

The Grand Duke bent a quizzical look upon Herr Weingartner.

'You are a sly dog,' he said. 'We begin to understand a number of things that have been none too clear. No doubt this lady is beautiful to look at as well as pleasant to hear?'

'Beauty is in the eye of the beholder, Highness,' Josef replied, dryly. 'The lady herself makes little claim to personal beauty. On her behalf I can say that she is . . . different.'

He fell silent for a few moments.

'Yes, quite different, certainly. Her like has not been seen in Strelsau for perhaps a generation.'

'You intrigue us, Herr Director Weingartner. How came this lady to be in Vienna?'

'The good "Friends of Music" of Strelsau gave their . . . support, so that she might study singing with Madame Marchesi in that city. Her name has not been much noised abroad, since she is still very young.'

'What is your definition of young?'

'Perhaps I meant, in the professional sense. In age, she will reach her twenty-first birthday only on the first day of the New Year.'

'Quite a coincidence, is it not, Herr Director? It appears that there are many things happening in this city of which I am unaware. Such a state of affairs must be remedied. I would also remind you that the idea of "young" is very much a matter of opinion.'

Josef's smile was bitter.

'Indeed, Your Highness. But we must also remember that even at so tender an age, there are those who can play, and those who must work.'

His Highness became thoughtful; the inference was very pointed. He was not ignorant of his own reputation in Strelsau.

'And when may we have the privilege of an introduction to this . . . lady?'

'Should Your Highness extend his gracious permission for the presentation of this opera, you will see her for the first time, on stage, on the evening of Your Highness's anniversary, at the precise hour of seven forty-five.'

His Highness smiled, irony upon his mouth.

'Until then, we presume she is to be held *incommunicada*. You are very discreet, Herr Weingartner.'

'One cannot be too careful, Highness. As for *incommunicada*, there is no such intention. The lady will not arrive from Vienna until late in the day before the performance, and a little rest will be necessary. Thereafter she will be fully occupied . . . until seven forty-five.'

He stressed the last words.

The Grand Duke made a gesture of mock frustration.

'It is understood. And the others? You spoke of professional artistes, did you not?'

'The Directorate are presently in communication with those whom they would wish to sing the piece,' Josef told him, a little mendaciously, since not all had yet been approached. 'It is hoped that the *tenore* Signor Giovanni Tortoni will play the Duke . . . .'

His Highness beamed.

'Truly a man after my own heart,' he declared.

'. . . Signor Demetrio Apostolos will play the Father Abbot; Signor Enrico Mentoni the Innkeeper and Signor Angelo Giacomotti his son. Madame Bianca Alvi will play the Abbess, a small part, and the Fräulein Isabella Weber will sing the role of *Cordelia*. There will be lust, betrayal, jealousy, and a fatal duel . . . .'

'Is there to be a happy ending, Herr Director?'

'Whoever heard of a happy ending in Grand Opera, Highness? But be that as it may, the singing and the playing of the parts, as I have seen and heard it done even by students, will be magnificent. Your Highness will have, in effect, your own opera, at its first performance, conducted by the composer himself, and sung by a wonderful soprano. Reflect, Your Highness, an all-embracing *première*, in your own city. Your name will be inscribed in the history of music as a most enlightened patron.'

The Grand Duke pondered upon this for some time, while Herr Weingartner waited, fearful that he might have gone too far.

'The cost of this gesture?' His Highness demanded abruptly.

Josef's spirits sank within him. He sought for appropriate words.

'For such a . . . memorial, Highness, no cost could be too great. I beg you to think on it . . .'

'The proposition is attractive — my personal merit apart,' the Grand Duke allowed, 'and your audacity persuasive, Herr Weingartner. For that you shall be rewarded. Your pupil, born here in Strelsau, shall sing in the first new offering of my continued interest.'

He slapped his thigh, and roared with sudden laughter.

'This will give my mum-faced Gentlemen and my feather-brained Ladies of the Court something to think about. Come, Herr Director, you shall see to it that I am publicly named as the particular Patron of this presentation and . . .' he fixed Herr Josef with a glittering eye, 'the protector of the Fräulein Isabella

Weber. The mouth of the Privy Purse shall be persuaded to open a little wider to support such a venture.'

Herr Josef Weingartner bowed.

'Your Highness is most generous,' he said.

But the sweets of success were even at that moment turning to ashes in his mouth.

*

The Opera House, normally a brilliant spectacle at this time of year, was in total darkness as they drove from the railway station into the Stanislausplatz in the direction of Isabella's modest home, the apartment of Papa Smith, but the Square itself resembled a Breughel picture of immoderation, gluttony, and licentiousness, of which the centre piece was a great fire which blazed almost under the nose of the Grand Duke Stanislaus's horse. Between the ears of this enduring animal a besotted underprivileged citizen had placed, after perilously scaling its bronze flanks, a red cap of Liberty. The Grand Duke himself stared straight in front, in frozen arrogance, but other Bacchanalian citizens danced about the monument to servitude, howling and screaming. The snow melted about the edges of the fire, and ran in dark rivulets reflecting the bloody glow in the interstices of the cobbles. Stalls had been erected wherever there was space; they were piled high with pies and cakes of every variety. Barrels, set up at the four corners of the square, dispensed free wine to the milling miasma of dark ghosts that swirled in from the surrounding mean streets to take advantage of this rare respite from human bondage. Lanterns hung from every projection of every house, and from the branches of the trees which lined the square, and under them, dark and fiercely moustachioed itinerant gypsies played wild Magyar melodies on their violins, eating and drinking whenever they could snatch a morsel or a cup. At the castle up on the hill an assembly was going forward in anticipation of the morrow; at this assembly music was provided by a select number of chosen volunteers from the orchestra of the Opera House.

The revellers paid little enough attention to the fiaker as it threaded its way with difficulty through the undulating mob. Maria and Christopher looked upon the scene with interest, but Isabella looked scarcely to right or left, her pale face as hard as that of the Grand Duke Stanislaus himself. There was, in fact, no little resemblance between the profiles of the two. She was in addition fatigued by the uncomfortable train journey from Vienna by the newly-opened railway line. She was indeed so weary that she was unable to appreciate the modest dinner arranged for the homecoming heroine by Papa Smith at his modest apartment. She had few words for anyone, and retired early.

The Doctor was condemnatory about the whole affair.

'She is in no condition to go through with this performance,' he told Herr Weingartner. 'The child is already exhausted. I warned you how it would be long ago. She needs a long rest, complete relaxation. Otherwise you will of a certainty lose your ''Nightingale''.'

Herr Weingartner was uneasy.

'It is the journey merely,' but he did not convince himself. 'She will be better for a night's sleep.'

'Tchah!' said the Doctor. 'But we shall see.'

Herr Weingartner presented himself early on the following morning, ostensibly the bearer of anniversary gifts, but in reality to assure himself of Isabella's continued well being. He was determined to be as diplomatic and ingratiating as was possible. She received him pleasantly enough, without affectation.

She thanked him for his presents, gave him a cup of coffee, and inquired touchingly after Matthild'. Was she in good health, and still able to care for her well-loved home?

'Matthild' is well enough,' Josef allowed, 'the good Doctor attends to that most zealously.'

From the uneasy silence that followed this brief exchange it was clear that they were both playing for time.

Greatly daring, Josef put his anxieties into words.

'You are well rested after your journey, Isabella?'

'I am well enough,' she told him. 'I have slept and, thanks to Maria, I have eaten. There is little else to be done.'

'You are quite ready for this evening?'

There was another long pause.

'We have had long discussions, Christopher and I. We are agreed that there is little necessity to remain in Strelsau. If we are to get on in our chosen professions, this city is scarcely the place to do it. The world is full of places where I can sing, to much better advantage. I have decided. I shall sing where I please, when I please. It will not be so difficult.'

Josef's face was ashen.

'But Isabella, this evening . . . .'

'I doubt if I shall ever see Strelsau again, much less sing here,' she murmured.

Josef sat violently on the nearest chair. This then was the crux. In spite of the passing years she had learned nothing, forgotten nothing. Under the momentary mellowness lay the old implacable perversity. He dreaded what she might say next.

'But, Isabella, this evening . . .' he was almost pleading now. 'You will sing this evening? It is all arranged. Even the orchestra has been trained to a hair, by no less a person than Herr Smith . . . .'

Isabella's eyes brightened.

'Dear Papa. And he is so very old now. How could you do this to him?

'He was more than willing, Fräulein. You have no idea how proud he is of you. He was very happy to do this . . . for you.'

'It is not possible for him to conduct the performance. The strain would be too great.'

'Indeed, Fräulein. Signor Andelli has now taken over.'

'That is interesting,' said Isabella thoughtfully. 'He has said not a word to me.'

'I told you it was all arranged, Fräulein.'

'You are devious, Herr Josef, are you not?'

Josef smirked. He was pleased with himself. 'Signor Andelli arrived some days ago. At this moment he is about to begin the final rehearsal.'

Isabella sat up in her chair with an angry movement. Her eyes blazed.

'How dare he, without me? How dare you . . .? It will be useless, a pantomime . . . I must go at once. Christopher, call a fiaker. Maria, my cloak. Should you wish to accompany me, Herr Josef Weingartner, you may. I will sing in your Opera House tonight, and you shall indeed hear how "your nightingale" can sing. But I would beg you to remember what became of that poor bird. Not,' she added maliciously, 'that you are an emperor.'

Josef ignored this remark, but his heart lifted. He endeavoured to express his happiness to the flying figure.

'You will not regret it, Fräulein,' he told her. 'To sing before Royalty, before the Grand Duke himself, the Patron of our musical endeavours, from whom, personally, we derive our support and encouragement. What an honour . . . .'

He became aware, in the stream of his reverie, that Isabella had stopped behaving like a whirlwind, and had fixed him with those cold green eyes which had so unaccountably demoralized him in the past.

'Herr Weingartner.'

The voice cut like a knife.

'F . . . Fräulein . . .?' he stuttered.

'The Grand Duke?'

'It is his anniversary, Fräulein, as you know. It is the reason for this festivity.'

Isabella made an angry gesture.

'The Grand Duke . . . Patron . . . .'

Once again Josef found himself in the familiar defensive position. He waited instinctively for the blow to fall upon his hapless head.

'Indeed, Fräulein.'

Isabella all but sneered in his face.

'*Patron*. Patron . . . . Patronage. *Gott in Himmel*. Is it not possible to succeed in this world unless one is propped up by some arrogant nobody, albeit a petty princeling? I owe no one a debt, not even you, my dear Herr Josef Sigismundus Weingartner,' (but her eyes softened momentarily as she said this). 'My voice is my own; you only built a little on what foundation already existed. You have said so many times. Why then should you, or any one else claim the credit? My voice would have become what it is had you never heard of me. The result might have been delayed a little longer. Yes, perhaps I do owe you a debt, Herr Josef. But this Grand Duke . . .' the scorn in her voice was bitter and cutting, 'this privileged *fainéant*, this sprig of lawless provincial Royalty — I assure you his name is a byword throughout Middle Europe — what have I to do with him, or he with me? Patron, faugh . . . .'

She could not make her disgust plain enough.

'You have gone too far, Herr Weingartner,' were her last words as she left the room.

*

That evening, in common with the rest of central Strelsau, the Opera House blazed with light. Light shone from every orifice within and without; it shone upon the grey trodden snow before the newly-scrubbed steps; it shone upon the streaming flanks of the horses and on the shining carriages as they drew up before those same steps; it shone upon the gleaming uniforms and the white bejewelled bosoms of those who considered themselves essential to the adequate social life of the Grand Duke; indeed, several of them were on more than adequate terms of acquaintance with him. Their presence at what was in effect a Command Performance was bound to be remarked. In carriage following carriage they came, the ladies and their gentlemen; two by two they ascended the steps; two by two they paused at the entrance to the auditorium and surveyed the House, and were surveyed by it in their turn, before they took their places with much ado about nothing to await with disingenuous interest the first appearance of the Grand Duke's latest enthusiasm, the Fräulein Isabella Weber, newly arrived from the Opera in Vienna. It was not often that such a situation ruffled the surface of life at Strelsau.

It was the more remarkable in that His Highness had not yet even met the lady, although it was rumoured that she was a native of the city. Not even the 'Friends of Music', when they were quizzed about the matter, could offer any information. Remarkably tight lipped, they were.

If Isabella was aware of the gossip about her that was by now common property she gave no sign. She was ignorant of the fact that His Highness had more or less let it be known that her forthcoming appearance was his personal triumph. She intended to treat the occasion for what it was, that is, part of her normal way of life as it would be lived from this time on. She entered the Stage Door, accompanied by Maria, at precisely seven of the clock. Maria went to the dressing room, while Isabella turned into the little office from which Hans kept a vigilant eye on the proceedings in the corridor. She sat in the familiar chair before the fire.

For some time nothing was said, then Hans looked at her over his half-spectacles, grunted, and turned again to his coffee making.

'The office is private, Madame,' he informed her over his shoulder.

'Hans, Hans, do you not know me now?' she reproached him.

He turned once more, and studied her critically.

'Madame?'

'I am the Fräulein Isabella, soprano, of the Viennese Opera.'

She rose and pirouetted before him.

There was a hint of mockery in his eyes.

'Good evening, Fräulein. Welcome to Strelsau, Fräulein.'

She drew herself up to her full height. Her burnished hair and her eyes shone in the firelight.

'You are teasing me, Hans. Or am I so changed?'

'I see little change, Fräulein,' he said. 'You are somewhat taller and, if I may say so without offence, a little plumper. You are better-looking than you were, quite handsome, in fact. You did not come to see me this morning as you passed by in great haste. Your face was stormy and unlovely.'

'I regret that I was angry, Hans, with good reason.'

'Indeed, Fräulein? There was, I remember thinking, some exhibition of mortification. But you are still the little Isabella who used to stand in the corner of the wings and listen to the singers, while Herr Smith played his fiddle down there. There were times when you fell asleep in that very chair. So innocent you looked then. So innocent.'

Isabella pretended to be annoyed.

'You *are* teasing me, Hans. You do not know me, after all.'

'If you think so, Fräulein.'

At that she embraced him warmly.

'I love you, Hans. It all seems so long ago, and many things have happened so that I am quite frightened sometimes. I have thought of you so often, in your little room here, making coffee, and greeting the singing ladies as they came in. Now it is my turn. Tonight I am to sing on that same stage. Before the Grand Duke, no less.'

He turned away from her to his coffee pot, grumbling to himself. 'The higher they rise, the faster they fall.' His memory was quite sharp.

'You said, Hans?'

'Nothing of importance, Fräulein. A passing thought, that was all.'

'I always wanted to sing, Hans. Papa is very pleased. He says that I sing quite beautifully.' She said it as a child might say it.

'Then it must be so,' Hans assured her, and his thoughts went back over the years. *It does not matter whether you sing well or ill; either way I see trouble.*

From across the square the great bell of St. Michael boomed its indifferent message of passing time above the noise of the revellers below. It was a quarter after the hour. The carriages were increasing in number, the drivers forcing their way through the wanton crowds with much shouting and hauling on the reins to achieve the appropriate slippery stop before the steps. The crowds closed in about them so that beauty and aristocracy might be more judiciously examined as it descended from those same carriages.

Isabella stirred as the sound of the bell penetrated into the office. Yet it appeared to Hans that she was reluctant to leave.

'It is time, Fräulein,' he said, clearing his throat loudly.

'As you say, Hans, it is time. But they will not seek me here, I think. Here it is warm and comfortable. Out there it is cold, unfriendly. I do not think that I wish to sing. It is too much, Hans . . . . It is too much.'

'Poor child,' he thought to himself. 'Too much, too soon . . . . Poor child.'

Aloud he said; 'Nevertheless, Fräulein . . . .'

'I know, Hans, I know. I have been taught to be fully professional. I must go on and sing, even though my heart is breaking.'

She made a mock of wiping a tear from her eye. Hans regarded her with a new curiosity.

'And is your heart breaking, Fräulein?'

She smiled, tremulously. The tears were real enough.

'Indeed not, Hans. Nor is it likely to be.'

She held her hand out to him. He took it, and pressed it gently for a few moments.

'Goodbye, Hans,' she murmured.

'God be with you, Isabella,' he replied.

*

Andelli stood on his rostrum, his arms hanging limply, his whole body perspiring. It was over, and he was almost physically exhausted after three hours of his own convoluted music. There had been passages during which he had experienced the effort of pushing a great rock forever uphill. Yet the orchestra had done itself credit for such a provincial body of musicians. They could scarcely be expected to have the touch, the expertise, the understanding, of the gentlemen of the *Musikverein*, and surely Isabella must have felt the difference. She had made no comment; indeed, she had remained behind the closed door of her dressing room throughout the intervals. In vain he, and Christopher, had begged admittance. They had been turned away ignominiously by Maria, acting upon her mistress's instructions.

In the Royal Box His Highness the Grand Duke, surrounded by the familiar, boisterous companions of his age both male and female, waited with mounting boredom and signs of impatience for the end of the performance. As usual he was more interested in the performers than in what they were playing. In particular he could not take his eyes from that proud figure as it stood at the final curtain looking down dispassionately upon an audience almost as abandoned within the House as the crowds had been earlier in the square outside. It stood, it cheered, it called her name in unison. It clapped endlessly. The uproar continued even when Isabella turned and walked off stage, and the lights went up in the dusty auditorium.

Nor did she return. The roar abated perceptibly, but when His Highness rose in his place, and looked fleetingly down upon his people, it began again, and the clapping sounded like the sea breaking upon the shore. Before it had begun to subside he was making his impetuous way backstage.

He threw the door open, to be confronted by the intimidating figure of Maria, and even he was disconcerted.

'The Fräulein Weber . . .?' he half-demanded.

Maria made a sullen curtsey, and with a sour edge to her voice announced him.

'His Highness The Grand Duke Karl of Strelsau, Fräulein.'

Isabella lay palely upon the chaise-longue. She neither spoke nor moved, and perhaps for the first time in his life the Grand Duke was bereft of words. When at last he found his tongue, his voice had lost its usual brashness. It was instead almost gentle.

'How beautiful you are, Fräulein. They had not told me the half.'

As he spoke the bell tolled again from the high tower of the Cathedral, scattering the roosting pigeons from the ledges. The snow continued to fall implacably, silently; until all vestiges of the day's jollification were covered beneath the winding sheet of winter. The city was silent.

It was eleven of the clock on the first day of the New Year.

*

The music and the dancing continued at the castle upon the hill. The dancers twirled and postured beneath the magnificent candelabra, languishing in each other's arms, their eyes and limbs heavy with sleep. Only the Grand Duke himself, it was observed, did not enter into the spirit of the affair with the same carelessness as he once had done. Twenty-one he might be, on this very day, or was it yesterday, but he was no longer the free-and-easy young man he had been, even only a few hours ago. He appeared withdrawn, almost unapproachable.

Down below, in her small room in Papa Smith's apartment, Isabella found sleep more and more evasive.

*

As the great bell reached its eleventh stroke on the morning of the following day, His Highness brought his horse to a spanking halt before the portal in the narrow street, and almost before he could descend from his sleigh that grey snow-bound alley was alive with shadowy forms hovering at a discreet distance from the equipage. They moved, mingled, coalesced in the narrow confines. They murmured, they whispered, but the confidences these ghosts exchanged were inaudible to any but themselves. As they waited Isabella appeared and stepped into the sleigh. The murmuring and the whispering of the ghosts became louder. Silence fell again as the Grand Duke whipped up his horse, and the party moved off in the direction of Zenda. The grey ghosts were not to know that it turned aside long before that place was reached.

Papa Smith, his face a mask of misery, sat with Christopher in his apartment. He had pleaded with Isabella; he had pointed out, in so far as he was able, and perhaps not in the most tactful of terms, the danger that lay in her path. He had, as a last resort, commanded that she should not associate with this libertine. That had been the last straw.

'I am my own mistress,' she had told him. 'I am of age, and my life is my own.'

There was nothing he could do. She was indeed her own mistress.

End of ACT II, Scene ii

INTERMISSION

INTERMISSION

On the third day, after Isabella had returned, without explanation for her absence, Herr Smith's resolution stiffened. Conscious that he had now very little authority, and would not indeed have attempted to impose it, he yet felt that he had some kind of jurisdiction by virtue of his pseudo-fatherhood over her comings and goings; with the knowledge that only he possessed, he was sure that the situation could be satisfactorily resolved.

He dressed himself with great care, and at some expense took a *fiaker*, the better to approach the Castle in as dignified a way as possible. Aware of his power, he had no intention of playing the suppliant.

In the cold reception chamber, whither he had been conducted by an ancient Lord Chamberlain, he stood face to face with arrogant royalty, and his resolve began to weaken.

For some time nothing was said. At last His Majesty spoke.

'Herr Smith?'

'Your Majesty.'

Herr Smith bowed.

'Of the Opera?'

Herr Smith bowed again.

'I have been Principal Violin for a number of years, Your Majesty.'

Again His Majesty stared at him.

'You are English, Herr Smith?'

'Indeed, your Majesty.'

'You are privileged, Herr Smith. We have musicians of our own. What then is the nature of your complaint? Are you not treated well in our country?'

'I have received the best of treatment, Your Majesty. I have no complaint.'

'Why, then, do you seek audience, Englishman?'

'There are matters of some importance, Sire, with which I feel you should be immediately acquainted.'

'They must indeed be of some importance, since I consider that I am well informed on all matters concerning my realm.'

Herr Smith struggled to find words that would counter the sensation of patronage; words that would revive his former decision to be the master of the situation.

'These matters concern both you and me, Your Majesty,' he said, rather diffidently.

A kind of smile distorted His Majesty's lips.

'It is my opinion that such a thing is most unlikely, Herr . . . Smith. Are we unknowingly related in some matter of importance?'

'Not related, Sire. Connected. Unwittingly and in my own case, as a consequence, unhappily.'

'Indeed, Herr Smith. And you do not relish this connection.'

'With respect, Sire, no.'

'You are at least forthright, Englishman. But then, so I am informed, are most of your countrymen. There was one, Rudolph, I remember . . . .' he broke off. The bantering tone became harsh; the features hardened.

'Your matters of importance, Herr Smith. Lay them before me.'

Herr Smith began, after some hesitation:

'With permission, Sire, I will return to some twenty-two years ago . . . .'

Sire stared steadfastly before him, his lips puffing gently in and out. But he said nothing.

'At that time, in this city,' continued Herr Smith, moistening his dry lips, 'there lodged a certain Fräulein Fanny Bamberger. She was, Your Majesty may or may not recollect, an actress at the *Komedie-Theater*.'

In silence His Majesty continued to push his lips in and out, and his eyes were closed. For a very long time the silence persisted. The Lord Chamberlain wrung his hands; an expression of anguish passed across his features.

'So?' His Majesty gave voice at last.

The Lord Chamberlain prodded Herr Smith gently in the ribs. 'Continue,' he urged the Englishman.

'Your Majesty may recollect also that the lady was given to wearing the finest clothes, parading the most valuable jewels, riding her own carriage, and entertaining on what can only be described as the grandest scale. She was young, she was beautiful, and she was certainly most indiscreet. She made little secret concerning the source of her comforts. Your Majesty may recollect that she was suddenly seen no more in Strelsau.'

His Majesty's eyes were narrow slits. His mouth assumed a vulpine smile, the lips stretched in a tight grimace over discoloured teeth. The rages of His Majesty were notorious; indeed, his physical reaction to persons or things that displeased him were apt to be violent. Herr Smith drew back a little, afraid that he might be the target of the King's reflexes.

Yet there were more important issues at stake than the displeasure of a despotic prince.

'You had a brother, Sire. An elder brother.'

His Majesty made a gesture of impatience. When at last he spoke it was with a great effort.

'So much everyone knows. And everyone knows, to anticipate your next offering, Herr Smith, that my brother died after a fall from his horse while hunting.'

The Lord Chamberlain crossed himself. It was quite unnecessary for him to say anything, but he was not distinguished for the strength of his intellect.

'That is so known,' he babbled. 'That is so known.'

'That is what your Majesty wished, and still wishes to be known,' Herr Smith corrected him gently. 'What is not so well known is that Your Majesty's brother had not ridden his horse for any reason on the day he was supposed to have died from an accident in the hunting field. In fact, owing to certain complicated problems of State not a little concerned with yourself, Sire, he had not ridden a horse for some time previously.'

The King's eyes narrowed again. His hands gripped the arms of his chair until the knuckles stood out deathly white. Herr Smith felt a fleeting sense of triumph; he had indeed drawn first blood. Now there was no going back. What he had come determined to say must finally be said.

'Your brother was, as Your Majesty only too clearly recollects, incarcerated in the Castle of Bromberg some time before that tragic fall, upon your father's orders. It has been rumoured that, so far from dying at least honourably on the hunting field, his unfortunate demise was hastened, certainly by a fall, a fall down the steps of the Castle during a drunken brawl. It has also been rumoured that this fall was expedited by a discreet push from one of those who were so disgracefully involved that day. It was the last of a series of disreputable episodes, Sire; a timely solution, or a well-managed one, to a problem beginning to weigh heavily upon certain of the House of Elphberg.'

The King's face was dark with anger.

'You are too familiar, Herr Smith, with the most intimate secrets of our House. This is an impertinence not to be borne. Were this another age, I would have you incarcerated also.'

'That I do not doubt, Sire. It is perhaps as well that we are sensible men. It was all so long ago. I had no wish to acquaint myself with the secrets of your House, but I was compelled to make certain inquiries on my own behalf. All roads lead to Rome, your Majesty, and those inquiries inevitably led me back to you. The knowledge revealed to me remains secret still and will, God willing, die with me.'

'That may be. That may be, Englishman. The matter shall be considered. Meanwhile I am in some confusion as to the manner in which inquiries for your own purposes could lead to the acquisition of such knowledge as you now possess.'

Herr Smith reflected that perhaps the worst was over. He was still at liberty, when it might have been otherwise. He felt increasingly bolder.

'We return now to the late unfortunate Fräulein Fanny Bamberger . . . .'

His Majesty loosed an indescribable sound.

'Some years ago,' Herr Smith continued, 'I made myself responsible for a waif of the streets of this city. The reason still defies explanation. Foolish sympathy, anger, sadness at man's inhumanity to man, all of these. Or perhaps the workings of Fate. The child was ill-dressed, ill-treated, unwashed, suffering from malnutrition. She was compelled to beg, to sell matches, and trinkets probably stolen, to satisfy the rapacity of her parents. That is, I understood that they were her parents . . . .'

'And they were not?'

'They were not, Sire. There lies the beginning of my story. I paid such money as I could spare to them, and took her away. They were in the end quite easily persuaded to be rid of her. They were, it appeared, afraid . . . .'

'Of what, Herr Smith?'

'That I do not know, Sire. Some belated retribution, some punishment, perhaps, if they said too much.'

'Who were these people? Their names?'

'With respect, Your Majesty, that is another secret. They did no wrong at all. They were paid enough for the little they did do. At least the child had a home of a kind, and they cared for her after their fashion. But I took her into my home. She has been a daughter to me.'

The King stirred uneasily in his chair.

'The charity is greatly to your credit, Herr Englishman,' he said. 'Would that all English were so-minded. We are grateful for your interest in one of our smaller unfortunate citizens. At this late date, however, I fail to perceive why I should be concerned, although I see that you expect to find me so. Your discourse lacks direction and cohesion.'

'Both qualities will appear in due course, Your Majesty.'

'To the point, then. To the point.'

Herr Smith was now in full control of the situation. He bowed slightly, and resumed his narrative.

'The child was a good child. She was friendly, obedient, amenable at first to kindly discipline. Regrettably, in the course of time, grave faults of character began to appear.'

A sudden glint of amusement appeared in the King's eyes. He considered Herr Smith with mock gravity.

'And these faults were, Herr Smith?'

'As she grew older she became exceedingly stiff-necked, wilful even.'

His Majesty actually chuckled.

'She had perhaps red hair,' he said. But it was a statement rather than a question, and Herr Smith was momentarily disconcerted.

'I have endeavoured by various and devious means to eradicate this obstinacy, this self-will derived from who knows what source. All has been useless. Yet I have not altogether failed in my ministration, for in general she has a lovable character, and her intellectual qualities are of the highest order.'

'Quite a parcel of inconsistencies, I perceive. I suspect, Herr Smith, that in finally praising the pupil, you begin to flatter the teacher.'

'Your Majesty, forgive me if . . . .'

'Quite inadvertently, no doubt. It is a human weakness. But continue, if you please. Through the fog of your incoherence a pattern begins to emerge.'

'The child is now a woman, Sire. You will, I am sure, forgive my feeling of satisfaction when I say that she is able to take her place with the highest.'

His Majesty raised his shaggy eyebrows.

'It is scarcely my doing,' Herr Smith protested with haste. 'I have been but the instrument, and as an instrument I am undoubtedly imperfect. There are many times when, in spite of my most earnest endeavours, her judgment on men and matters is lamentably weak.'

'Lack of judgment and self-will combined are ever difficult to deal with,' His Majesty remarked. 'Am I to understand then that this lady — fallible in her judgment of men — has committed some prime folly, and that you require my help in the matter? It is scarcely within the scope of my duty to . . . .'

'Your pardon, Sire. The *lady*,' and he emphasised the word, 'has had the misfortune to attract the attention of Your Majesty's son, the Grand Duke Karl.'

'Be careful in your choice of words, Herr Smith.'

'Your Majesty, to me the situation can result in nothing but misfortune. I use the word advisedly. It occurred to me that in view of past events Your Majesty might see it in the same light, especially as there is evidence that the lady returns his interest.'

There was silence for a moment or two; then a smile appeared slowly on His Majesty's face.

'It is a familiar theme, is it not? A turn in the plot so much employed by the writers of romantic fiction. I have no doubt it will be but temporary, Herr Smith. It has happened several times already, the Grand Duke is a hot-blooded and susceptible young man.'

The Englishman's face darkened.

'Your Majesty is pleased to treat the matter lightly. It is no matter for jesting.'

'The name of this paragon among women?' the King demanded.

'The lady's name, Sire, is Isabella Weber.'

The King combed his beard with long fingers. His eyes were narrow, and his mouth tight. He considered what this Englishman might want, in his righteous anger. The situation was not without precedent. Perhaps some recompense for loss of . . . dignity?

'Isabella Weber. So . . . . But I do not understand, Herr Smith, your previous references to my late lamented brother and his — ah — escapades. I should appreciate further elucidation.'

Herr Smith held forward a small square of paper. His Majesty peered at it from a distance.

'Approach, sir,' urged the Lord Chamberlain. 'Approach His Majesty. His sight is somewhat dimmed.'

'Silence!' The voice was thunderous. 'Bring me that paper.'

The Lord Chamberlain took it delicately between thumb and forefinger, approached His Majesty, and held it at the distance of six inches from the royal nose.

Herr Smith coughed gently.

'Your Majesty will no doubt recognize the features of Your Majesty's late elder brother, the Count Rupert. And this,' Herr Smith held out a similar square of paper, rather newer than the previous piece, 'is a portrait of the Fräulein Isabella Weber.'

In silence the King took the portrait. His fingers trembled as he compared the two.

He held the portrait of the Count Rupert forward.

'How did you come by this?'

'It was given to me by the woman who was known as the child's mother, under a vow of secrecy. It was indeed the only proof of identity the child had. I have preserved it with the greatest care, scarcely thinking that one day it would be of importance. If Your Majesty has not yet arrived at the solution to this problem . . . .'

'God's death, man. I am not an idiot.'

'The child was the daughter of Fanny Bamberger; her father was Your Majesty's late brother, the Count Rupert. In short, Sire, the Fräulein Isabella Weber, *Prima Donna* of the Vienna Opera, is your niece, and the cousin of your son the Grand Duke Karl Ernst Maria Michael Frederik.'

A kind of groan escaped the lips of the Lord Chamberlain. He waved his hands in a distracted fashion.

Ignoring him, Herr Smith continued.

'You will perceive, Your Majesty, how impracticable even a temporary romantic association between these two must be.'

'Is it so impracticable?' the King replied, with a twisted smile. 'They are but children yet. Far less practical liaisons have been contracted in the course of history. At least one knows their line of descent.'

'They are blood cousins, Sire.'

'So . . .?'

'The strictures of Holy Writ, Sire . . . . Indeed, it may already be too late.'

In a heavy silence they stared at one another, two old men confronted by the ghosts of the past and the uncertainties of the present. And as for the future . . . .

'It cannot be allowed to go on, Your Majesty. At the very least it might be that the Fräulein's career would be shattered.'

'I am less concerned with the Fräulein's career than with the name and status of the Royal House.'

'If this were by any chance to come out,' said Herr Smith, 'the consequences would be little short of disaster. To begin with, the whole story would inevitably be subjected to a close investigation by the writers of the daily chronicles.'

'Those damnable newspapers,' His Majesty exploded.

'It cannot be allowed to go on,' Herr Smith urged. 'Separate them before it is far too late. Let them go their ways, sorrowing perhaps, but happily in ignorance of their relationship.'

The King's features were hard and grim. He spoke as though the words were being forced from him.

'You leave me no alternative, Herr Smith. You have gained your point. They shall be so separated. You have my word on it.'

The Englishman bowed.

'I thank you, Your Majesty, for your understanding.'

But he felt no sense of triumph, rather a deep sadness.

Some days later a large white envelope bearing a purple seal was delivered by special Palace Messenger to Fräulein Isabella Weber, at the Opera House, Strelsau.

The enclosure was brief, without preamble or explanation.

> '. . . In so far that you, the superscribed Isabella Weber, have by attested account, seduced the Grand Duke Karl Ernst Maria Michael Frederik, Heir to the Throne of this Realm of Ruritania, into amorous liaison, you are therefore herewith commanded to depart, with such possessions as you may personally carry, beyond the confines of the said Realm of Ruritania, and you are forbidden to return thereunto and therein during our lifetime.
>
> 'Given under our seal, this Twelfth Day of January, in the Year One Thousand Eight Hundred and Seventy Seven . . . .'

On the same day Herr Smith received a communication, plain and without a seal, and subscribed by the Lord Chamberlain. It too was brief:

> 'It would be advisable, Herr Smith, if you were to remove yourself also beyond the confines of our Realm.'

*

ACT III, Scene i

Outside the Nunnery of the Madonna dell'Angeli. Two days later

LONDON — STRELSAU.

ACT III, Scene i: Outside the Nunnery of the Madonna dell'Angeli. Two days later.

LONDON — STRELSAU

From the Grand Duke himself there came no word.

Day after day Isabella sat mutely by the window, staring unseeing into the dismal street below, waiting, listening for the urgent beat of horses' hooves, the hiss of sleigh runners or perhaps the rolling of wheels over the muting snow.

His words — the last words she ever heard from him — *I will call for you again, tomorrow*, burned in her memory and on her heart. But he had not come. All through that 'tomorrow' she had waited with happy expectancy; the following day with disappointment, seeking excuses for his lack of courtesy, and finally with anger. The words of Papa Smith's warning tormented her also; how much wiser he had proved to be than she, in her innocent arrogance, had thought him. Now, how impossible it was that she could think of confiding in him. She could not even confide in Maria, who nevertheless seemed to have a full understanding of the situation.

There was nothing to do except to obey the horrendous imperative, which she read over and over again in an endeavour to solve the mystery. The more she read, the less she understood. She, Isabella Weber, a child a mere twenty-one years old by a matter of a handful of days, in amorous liaison with the heir to the throne of Ruritania. 'Seduced the Grand Duke'? What sin had *she* committed that it could be described in such words? How big a stone was necessary to kill a butterfly?

His own secret locked in his tormented mind, Papa Smith was in no position to offer either explanation or even consolation. The folly of his impetuosity was only too obvious. He had not expected things to come to this pass. Who could ever predict events that would follow upon the fortuitous but fateful conjunction of a lusty royal scion, a gifted but parentless opera singer, a jealous adoptive father, and a tyrant of a king, with a background of music, and set in such a country as Ruritania? Could it be considered dispassionately, the situation had all the elements of *opera-bouffe*. Papa Smith was no Mozart; he could not resolve the tragedy he had set up with a happy quartet. Yet he consoled himself with the thought that Isabella's life was her own, as she had said. She was responsible, as she claimed, for her own actions. If she persisted in refusing to consider the possible consequences, so be it. He could have wished her a little less self-willed, a little more humble. There was nothing he could do; it is the nature of human beings to act first, and regret the action thereafter.

'What shall I do, Papa? What shall I do?' Her voice was the voice of despair. 'I do not care to live.'

His own face was wretched and woebegone as he endeavoured to comfort the weeping girl. He raised her up from where she had fallen across his knee, and essayed a twisted smile. Almost without knowing it he spoke to her as the child she once was.

'There is still much to do, *Liebchen*. The world is still not at an end, however much you may feel that it is. You can still sing . . . .' (but even as he said it he was fearful of the result such a shock might have had). 'Do you not remember how you wished to sing like Jenny Lind, so that people would clap and give you money? You were a child then, Isabella, with childish dreams, but you could always sing. Now you sing better than Jenny Lind ever sang. Does not that mean anything?'

So he sought to comfort the child, and diminish his feeling of guilt.

Isabella turned her face up to him, tear-stained and miserable.

'I do not wish to sing. I cannot sing, Papa.'

'*Liebchen*,' said Herr Smith, gently. 'All things pass. You cannot help but sing.' He gathered some strength from the sound of his own voice. 'I recollect that you have still three performances under Herr Richter. So you will return to Vienna and finish your contract. After Vienna, who knows? The world is more than Vienna. For you, there are no frontiers.'

'But we must live somewhere, Papa. Where shall we live, since we cannot live in Strelsau?'

His eyes closed; his mouth trembled.

'It matters little "where". "Why" is more to the point, Isabella. As I have said, for you there are no frontiers. Music is international. For my part,' and here he opened his eyes and looked into hers, seeking some confirmation of the decision he was about to make, 'I am growing old, child. I no longer have the desire to continue the life I have been living. Music is no longer for me what it was. It was changing beyond my ability to understand it. Even your opera, wonderful though it is, is scarcely to my taste. Time moves on, Isabella.'

'You are talking nonsense, Papa.' She attempted a tearful bantering tone, taking his hand in hers. They were like two children comforting one another, yet each unable to understand the troubles of the other. 'You are young enough to have rehearsed my opera, as you call it, whether you understood it or not. One cannot grow old, loving music.'

He made no reply, but sat stroking her bronze hair, staring the while into the distance at the lower purple flanks of the southern mountains and their cold white heads silhouetted against the grey sky, only just visible above the snow-covered roofs of the lower town.

'My work here, such as it has been, is done. You have achieved what I hoped you would achieve, Isabella, and I have done what I set out to do in helping you achieve it. So, my work is done. You have been a loving, lovable daughter to me, Isabella. For you I thank God. Christopher is an honourable and trustworthy son. It has long been my hope that you and he might one day marry.'

He fell silent again, and continued to stroke her hair.

'Christopher has taken the best care of me,' she acknowledged, not realising the irony implicit in the statement. 'I love him, like a brother, but I had not thought of marriage . . . .'

The tears fell again.

'I was a foolish old man, dreaming foolish dreams, child. Old men are well known for it, and the young rarely confide their intentions to such as I. They have their own lives to live, after all.'

'I am so very sorry, Papa, to have disappointed you.'

He gave her a sharp look. He could not remember any situation in which Isabella had admitted that she was at fault.

After some time he went on: 'That being so, there is little to keep me here in Strelsau. The memory of my dear wife will be with me wherever I go. She was a good wife to me during the short time God allowed her. She will understand if I leave her in her native earth in the cemetery on the hill yonder.'

He paused again, and thought his own thoughts.

'She would, I think, have been proud of you, Christopher,' he continued, addressing his son who was sitting miserably by the stove, at a complete loss to understand the situation. 'Her dearest wish was for a tall strong son, and perhaps thereafter for a daughter. Well, she had a son . . . .'

His voice changed suddenly. He thrust Isabella away, but gently. She went over to Christopher and, kneeling by his side, put her arm through his in a sisterly gesture.

'Dear Christopher,' she murmured, 'my dear brother.'

'I have decided,' Papa Smith announced in a firm voice, 'I will return to England. I will have a cottage in some small village where there will be trees and grass and flowers. There must above all be grass, soft English grass such as you have never seen, Isabella. Grass is the most beautiful aspect of all in England.'

'What has Isabella — or I, for that matter — to do with grass?' Christopher demanded. 'Isabella needs music. There will be no music in such a place as you describe, Papa. There is indeed little music in England, at all.'

'There is indeed music in England,' Papa Smith defended his country with some spirit. 'Herr Handel has seen to that. Mozart visited England, and found a welcome, although,' he reflected, 'he was but a child. Even your Hans Richter, Isabella, is about to go to England . . . .'

'One swallow — or even two — do not make a summer,' Christopher protested. 'A retirement in England may be the best for you, Papa, but what is to become of us, Isabella and me?'

Herr Smith considered his son in silence for some time. When he replied there was a suspicion of bitterness in his tone, tinged with resignation.

'You will no doubt do as your inclination guides you and as opportunity presents itself. You are both of an age to determine your own destinies, it is clear. You, Christopher, have had the best education that I could pay for, and you are an accomplished violinist, although I have seen no signs as yet of your setting the world on fire with your virtuosity. As for you, Isabella, the world of music has already heard of you. It will hear more. There is no doubt of that. You are both at the beginning of your lives and your careers. I do not think that there will be any lack of employment for either of you, capricious and treacherous though the world of music is. I see no limit to your future.'

With that Christopher had to be satisfied.

As far as Isabella was concerned the future did indeed seem to be without limit. Assuming the initiative, it directed its steps to the door of the modest house the four of them — for Maria was now to all intents and purposes a member of the family — acquired in the developing area behind Wigmore Street, in the Parish of St. Pancras. There was still, in this area of the expanding city, some vestige of village life, and still some green grass to be found in unexpected corners, but it was far from the dream of the return to his native country with which Papa Smith had sought to comfort himself. His health deteriorated steadily, and he took little interest in what went on around him, being increasingly dependent upon Isabella for the assuagement of his daily miseries. She, in her turn, acquired responsibilities the force of which she would not have recognised a few months before, burdened as she now was with household and family cares. She had the help of Maria, of course, but Maria herself was an unsophisticated child in a strange country, and little use outside the house.

It was to this establishment, therefore, that the future announced its incipient presence, through the medium of the Penny Post, in the form of propositions either supplicatory or even somewhat mandatory in tone, that Fräulein Weber should sing with this or that orchestra, or should grace with her presence this or that concert hall. It was intimated that she could, within reason, name her own fee.

For a long time Isabella resisted the seduction of these propositions, being fearful of the reception she might receive from the strange English. But necessity, particularly the necessity of having to pay one's way, knows nothing of prolonged indecision, and consequently it was London, the capital city of a land given over to strange folk songs, incomprehensible sea shanties and tasteless music-hall ballads, the London cruelly derided by Herr Hans Richter, that experienced the first demonstration of the voice which, until then, had been the subject of rumour. The voice of Isabella was to drive from the minds of the London musical humbugs all recollection of the divine Jenny Lind.

It was inevitable that Herr Richter, descending opportunely upon this great 'Wen', should seek out Isabella. It was just as inevitable that he should be of a mind, fresh as he was from his year at Bayreuth and his recent production of *Die Walkure* at Vienna, to publicise his connection with the music of Herr Wagner and his worship of the Maestro who was much underestimated in this unmusical city.

It was with some reluctance that Isabella allowed Herr Richter to persuade her to sing at the recently completed Albert Hall in a series of concerts under his direction designed specifically to present the genius of Herr Wagner to the people of London. Isabella was not at all in favour; she remembered all too well the strictures of Hans the Doorkeeper back in Strelsau concerning the new music coming out of Bavaria: *A lot of noise . . . too much wind, one way or another* . . . he had grumbled; nor did the remembrance of Herr Weingartner's dislike of Wagner help her indecision. She had no desire even to attempt to produce unlovely sounds. Out of respect for her prejudice, Herr Richter chose the *Fünf Gedichte*, and for a subsequent performance, the *Liebestod.*

The concerts were a great success. The singer was personally a greater success. No 'Wagnerian' soprano the like of Isabella Weber had ever been heard or seen in London. She was so incredibly young, so breathtakingly beautiful, and her voice was able to melt the most perverse heart. Fashionable music cognoscenti found immediately a new idol, while those who had neglected to attend that first long-to-be-remembered evening, chiefly because of their indecision as to the future worth of Herr Wagner's music offerings, found themselves quite out of touch. Unable to recount their personal experience of the epoch-making performance they were compelled to draw what satisfaction they could from some two or three column inches of English understatement under the heading 'MUSICAL INTELLIGENCE'.

> '*Herr Richter at the Albert Hall*,' said *The Times* newspaper in its well-bred voice. '*Music-lovers were last evening pleasantly entertained at a concert consisting of a selection from the works of Herr WAGNER*. The performance was directed by the Austro-Hungarian *Maestro Herr HANS RICHTER*, the outstanding exponent of the works *of WAGNER, who has lately taken up* temporary residence in this city. *The vocal items* of the programme, *Five Songs for a Female Voice (Wesendonck Lieder)*, were interpreted by a charming newcomer to the *London scene*, one Fräulein Isabella WEBER. Fräulein WEBER, a native *of the city of Strelsau*, we understand, sings knowledgeably, and has *a more pleasing* presence than any singer we recollect to have seen *in a long time . . . .*'

The reader of these notes was intended to understand by them that Fräulein Weber was a foreigner, as distinct from English singers, was handsomely young, and could sing passably well. It was enough. At her second performance with Herr Richter, when she sang the *Liebestod*, London was in no doubt. There was not a vacant seat in the House, and the consequent applause was as politely enthusiastic as the English temperament allowed.

The cutting from *The Times* occupied a place of honour in the centre of a single page of Isabella's '*Memory Book*'. The restrained style of the report was in keeping with her own temperament, in that she herself had already begun to deprecate any form of ovation or untoward expression of triumph directed towards her. At all times she refused to discuss what she called her 'work', while those who braved the purlieus to the north of the Oxford Road and

Tyburn to reach her door were turned away without satisfaction by Maria. This attitude caused a great deal of gossip, as may be imagined.

There were also in the '*Memory Book*' two other extracts of some importance. One of these was from a news-sheet called '*The Star*', written by a person using the pseudonym of *Corno di Bassetto*, who demanded with some passion why singers did not, could not sing like the Fräulein Isabella Weber. Why, asked this person, did they not go to drink the waters of Strelsau?

They did not of course take this advice, since *Corno di Bassetto* was considered to be a person of arrogant and idiosyncratic opinions. The consequence was that Isabella quickly became the toast of the town, and Ruritania the fashionable place to be able to talk about to one's friends, giving the impression that one's grandfather, or one's cousin twice removed was quite familiar with that country.

For some time Isabella's engagement book was quite full, but as the season drew to a close, and the days lengthened into weeks, and the weeks into the warmer months of spring, the tally of her engagements grew thinner and less demanding. Isabella herself became quieter and more withdrawn. She conversed only with Maria in the peculiar German dialect which was her childhood tongue. To Christopher she said very little indeed. She was quite unable to discuss with him the degree of lassitude and the languor which possessed her increasingly, or the discomfort her swelling body was causing her.

There was very little she could say to Papa Smith, either. In March he was taken with a severe pneumonia. After a week of painful lingering between life and death, that gentle soul quietly gave up the struggle. He had not even seen his longed-for English grass, except as he had glimpsed it through the murky windows of the train in which they had travelled uncomfortably through the fields of Kent from Dover to London. Aware that she had failed him in many ways, Isabella bade him a sorrowful goodbye in the cemetery at Highgate.

Thereafter the house seemed strangely empty and cold. That which kept the 'family' together was gone; nothing was quite able to renew the bond that Papa Smith had forged between them. Christopher walked the streets of London for long morning hours, endeavouring, without much success, it seemed, to find some kind of musical employment. Maria was more than happy to take up her routine of cleaning and polishing and preparing meals. When not engaged in these pursuits, she gave her time to Isabella; dresses were discussed, arranged and re-arranged; the *toilette* was attended to; in particular the bronze hair was brushed endlessly and relentlessly until it shone again, and then dressed in the latest mode. When these preparations were complete, Isabella would issue forth to be helped into her carriage and, attended by Christopher, would take her afternoon drive through the more fashionable parts of Town, descending now and then to indulge in some discreet feminine shopping. There was no singing in these days, but Fräulein Isabella was by no means forgotten. Whatever route the carriage took there were those who recognised her, and acknowledged her with a little curtsey, or a lifting of the hat, as if she were royalty itself.

'There is the Fräulein Weber,' they said to one another, and to those who were not sure, as the carriage passed. 'Did you not hear her sing at the Albert Hall? My dear, her singing is quite out of this world. Such a striking woman too, quite "*épatante*".'

She gave a slight inclination of the head to those who acknowledged her, and withdrew into her own world.

On one such afternoon, more troubled in mind and body than usual, Isabella lay back in her seat, content merely to savour the late spring sunshine. Her eyes were half-closed. Christopher, with nothing of great interest to occupy him, glanced negligently at *The Times* newspaper lying on the seat beside him. He sat up suddenly, and gave a low exclamation of surprise.

After some hesitation, he addressed Isabella.

'There is news . . . from Strelsau,' he told her.

Isabella's eyes closed, firmly. She might not have heard.

'Isabella,' he repeated, 'I said there is news . . . .'

Her eyes remained closed.

'I heard you, Christopher. Is it your opinion that I should show some interest?'

'That I cannot say. It is certainly of some interest to me, considering the connections we English have with that country, even though such connections have largely entered the realm of fiction now. The adventures of Colonel Rassendyll, for example . . . .'

Isabella opened her eyes.

'Christopher, you are talking nonsense. We have nothing to do with Colonel Rassendyll.'

'Perhaps more than you think,' Christopher remarked darkly. 'Much more than you think, from what I have heard. The present Lady Rassendyll, who is not known for her reticence, is quite proud of her Ruritanian connection. The subject is scarcely a closed book, as you very well know.'

Lady Rassendyll was one of those notorious London hostesses whose inability to remain silent was fuelled by her great store of reminiscences concerning her interesting and quixotic ancestor. Her unrestrained loquacity was a permanent source of amusing disbelief to her friends and acquaintances, although her claim to know more about the darker side of life among Ruritanian royalty was thought to have some foundation in fact. Isabella was bored by her, and endeavoured to avoid her. But when other friends of Lady Rassendyll sought to relieve their curiosity by quizzing Isabella about her own connection with Ruritania, and any possible relationship between herself and the Rassendylls and the Court of that country, she was driven to the verge of fury. They were quite shocked when she asked them in her turn about their possible connection with Macbeth and the English Royal Family.

'I really do not know what they are talking about, or what I am supposed to be able to tell them,' she had complained to Christopher. 'What should I know? How should I know it? The English are such fools.'

'They are like all other fools elsewhere, Isabella,' he had replied. 'Pay no attention.'

But now these foolish thoughts were disturbed by 'News from Strelsau'.

'The news, Christopher?'

He stared at her blankly for a moment, until she poked petulantly with the ferrule of her parasol at the newspaper he still held in his hand.

'Oh, ah, yes. The news . . . . Do you wish that I should read it to you? There is only a short paragraph, Isabella.'

'Read it. Read it . . . .' Her voice was faint, as if she had by some means divined already the matter of the 'News'.

Through the warm afternoon of that last day of July, in the year 1877, Christopher read to her the account set down under the headline '*Intelligence from abroad*'.

The paragraph was to be the third of those important cuttings pasted into the *Memory Book*.

He read it clearly and deliberately, glancing at her from time to time. It was certainly a short paragraph, and quite to the point, for not even *The Times* considered it worthy of longer comment.

> 30th. July. Intelligence from a correspondent in Upper Austria reports that in the first week of the present month His Highness Karl Ernst Maria Michael Frederik, 5th. Grand Duke of Ruritania and Prince of Slavia, Count of Styrinthia and Illyria, Margrave of Karonia, Lord of the Vldava Mark, Grand Master of the Order of St. Michael, was united in marriage with Her Highness The Lady Amelia Viktoria Konstanzia Camilla, Princess of Zell.

An additional line of 'Intelligence' stated that the ceremony was the occasion of almost unbridled rejoicing lasting for seven days and nights.

When he had read it Christopher paused, expecting some sharp comment from her. When none was forthcoming, and he could bear her silence no longer, he said: 'Have you nothing to say, Isabella?'

Her features were pale and hard.

'What is there to say? What do you suggest I say? Shall I convey by means of the Penny Post my best wishes? Shall I congratulate him on his marriage to the Princess of Zell who, it is well known, is stupid to the n–th degree? Is that what you would have me do?'

Christopher was non-plussed. He had in fact no idea what she could do or say, if indeed there were anything to do or say. But he had not expected quite such an outburst.

'There are not many people of Ruritania in London,' he ventured. 'The best efforts of the Penny Post will reach Strelsau only with the greatest difficulty, I imagine. But there is the Embassy, and a short message of goodwill . . . .'

'A message of goodwill? God in Heaven, Christopher, what are you suggesting? I have been exiled from my country. I have been parted from the scenes and friends of my youth. I am about to bear a bastard child, your father has just died almost certainly of a broken heart, and you think I should send a message of goodwill.'

She was quite furious. She closed here eyes, and sank back exhausted into her cushions.

'You are indeed a fool, Christopher,' she told him, 'but I have been a bigger fool.'

Suddenly she sat violently upright, and gave a scream so loud that others taking the afternoon air turned fearfully to ascertain the reason for this unexpected commotion.

Isabella's features were ashen and twisted; her eyes were wide with pain.

'Home, Christopher,' she gasped. 'Take me home, quickly.'

Almost before Christopher had taken in the situation, the coachman had whipped up his horses, and the equipage dashed along the road in a cloud of dust. To Christopher, holding the trembling, perspiring Isabella in his arms, the short-enough journey was a nightmare. He chafed her hands; he wiped the perspiration gently from her brow; he uttered what he intended to be comforting words; they were in effect complete nonsense.

As though she had foreseen the situation, Maria was waiting at the door. Her expression reflected her emotions as she ran to meet the carriage which drew up at her very feet with a great jingling of harness and a lather of horses. Christopher leapt out almost before they had stopped and, one hand on the side of the carriage, the other held out, he waited to help Isabella down. She half-stepped, half-fell in her agony and, scarcely knowing what she did, slammed the carriage door behind her. It closed with crushing violence upon Christopher's hand, and fell open again. A great cry burst from Christopher's lips. He stood stupified, staring at his mangled hand, now shapeless and bleeding, and fell unconscious in the road.

'Get the girl indoors,' the coachman shouted to Maria. 'I will attend to this gentleman.'

The child was born that same evening, an object puny enough to arouse all Maria's compassion, and to become immediately the centre of her world, since Isabella, although exhausted and lying between sleeping and waking for some hours was, as it is said, 'Doing well'. She continued to progress satisfactorily, even to the extent of being able to smile as Maria attended to the child's needs from hour to hour, and crooned the while in her peculiar dialect.

'*Mein klein' s Madel' n*,' she intoned over and over again. '*Wie Du so schon' bist*.'

If it was not music to Isabella's ears, it was at least a sign that Maria had accepted the situation.

Christopher, having been lifted back into the carriage by the coachman with the help of a kindly passing gentleman to seek the attention of a doctor, now sat through the day in pain, nursing his damaged and useless hand. It was only after some days, after some visits by the doctor to the stricken household, and the pain had begun to wear off, that Christopher climbed the stairs to Isabella's room and begged to be allowed to enter.

He was shocked to see how wasted she was again, so that his memory went immediately back through the years when she had lain in similar fashion in Fräulein Weingartner's great bed.

He bent over her and kissed her.

'Isabella,' he said. 'Oh, Isabella.'

'Oh, Christopher,' she echoed, smiling wanly. 'We are such a futile pair, are we not? How is your poor hand? Maria has told me all. I did not know what I was doing. But I am so sorry, so very sorry, Christopher.'

She took the hand in its bundle of bandages, and caressed it gently. As he flinched a little, she said quickly: 'Is it . . . is it very painful?'

'Painful enough,' he told her, with a wry twist of the lips.

'It will recover? Tell me it will recover.'

'It will recover somewhat, but . . . .'

'But what, Christopher?'

'The doctor holds out no great hope that I shall be able to use it again.'

Horror was frozen on her face.

'Christopher! Your violin . . .!'

'There is nothing to be done, Isabella. It was an accident. You are not to be blamed, nor are you to blame yourself. In any event, there are other things to do than playing the fiddle. To tell you the truth, I was never greatly taken with it. It is not in my nature. I should never have made anything of myself. My father was correct in his opinion of that.'

She held her arms out to him, and her eyes filled with tears. Supported by his one good arm she wept bitterly for a long time.

'I will make amends, Christopher,' she said, when the tears were at last dry. 'You shall be the controller of the household, my man of business. After all, there are still four of us.'

She turned to the child, lying peacefully in its cradle by the bedside.

Christopher surveyed the crinkled mask and the bubbling froth about the lips with some distaste. Isabella watched him with some amusement.

'We are more attractive when we are adult, are we not?' she asked him, with a grimace of amusement.

'I have heard it said that babies are beautiful,' he said doubtfully, surveying the simulacrum of a human being, its fine mist of hair already tinged with an underlying copper sheen.

'It is a trick of Nature,' Isabella said, 'to persuade mothers to think so. Otherwise there would be very few women who would consider the business worthwhile. The tigress does not think her cub beautiful; the cow is not besotted with the beauty of her calf. Their animal nature recognises simply that there is a pathetic and, for a time, useless little creature that is dependent upon them. I do not think she is beautiful, although Maria says she is. I shall call her Lisa. Later I shall teach her to sing, if it is possible. Maria shall teach her to cook, and you, Christopher, shall tell her fairy tales. You shall be in the place of the father she is never to know.'

'Some time she will ask, Isabella. There will come a time when, perhaps, she will be entitled to know.'

'When that time comes, it will be time enough,' Isabella interrupted him. 'I shall think of something.'

End of Act III, Scene i

ACT III, Scene ii

Outside the Nunnery of the Madonna dell'Angeli. Two days later.

LONDON — STRELSAU.

ACT III, Scene ii: Outside the Nunnery of the Madonna dell'Angeli. The same place and time.

The news of the old king's demise found its place, naturally enough, in *The Times* newspaper. Somewhat restrained in tone, as befitted such an event, it was accompanied by further intelligence — '*From a correspondent in Upper Austria*' — announcing the forthcoming Coronation of His Highness Karl Ernst Maria Michael Frederik, Grand Duke of Ruritania, etc., etc.

The news caused little agitation at the breakfast tables of London, or elsewhere in England, for that matter, except in two households widely separated by social distinctions. Lady Rassendyll, having perused the paragraph, laid down her newspaper with mixed feelings. Among the many thoughts uppermost in her mind was the possibility of a protracted visit to Ruritania and its capital city; the chance at long last to distinguish with her presence those places which until now had been only the stuff of legend. She began to make plans.

In the comfortable breakfast room of the house behind the new imposing row of Wigmore Street, Christopher read his *Times* in studied silence before passing it to Isabella without comment. Lisa sat in surprised apprehension at this departure from routine, while Maria's eyes were wary and watchful.

Isabella read the paragraph as if she were mutely spelling out every word. Even Christopher regarded her with some fear in his heart. He was not prepared for what followed.

'At last,' Isabella said, quite gently. 'At last. For twenty years I have waited for this. At last one of *them* has now gone into permanent exile.'

Christopher half-rose from his place at the table.

'Isabella . . .' he protested. 'This is not like you.'

Her eyes blazed at him with green fire. A grim smile curved her lips. For a moment she confronted him, until with a sudden movement she thrust her chair from her, and left the room. Lisa, unable to understand the lowering emotional temperature in the normally warm and responsive ambience of her home, would have followed her.

'Stay, child,' Maria bade her.

'She may be ill, Maria.'

'Perhaps a headache merely. There is nothing you can do at this moment. She is best left alone.'

'She is not ill. Overcome rather by a certain emotion of which you have no comprehension,' Christopher added. 'I think it would be as well if you were to assist Maria in her kitchen this morning, Lisa.'

'In the kitchen?' Lisa put on her prettiest pout. It would have moved a heart harder than Christopher's. How like her mother she was, he thought, recalling the image of a child at a window in a slum street in Strelsau, teasing him because of his Englishness.

'Certainly. And you have my permission to refrain from singing for the rest of the day.'

Lisa opened her eyes wide.

'I have my reasons. You are not at College today. A little rest will not harm your voice. But you can still be a useful member of Society.'

She rose from her chair. 'Very well,' she said, and made a face at him, which brought a good-natured rebuke from Maria.

Lisa curtseyed in Maria's direction, and kissed Christopher.

'*Mein klein's Mädel'n*,' Maria murmured fondly as Lisa left the room.

'She is no longer a child, Maria,' said Christopher. 'She is quite twenty years of age, and she can behave quite well.'

'That is so,' agreed Maria. 'She only looks like her mother.'

She changed the subject, suddenly aware that her remark was open to misinterpretation. Folding the newspaper, she said:

'Do you suppose that this will make any difference, Herr Christopher?'

'Who knows, Maria, who knows? It is too late, I think. Far too late. And yet . . . .'

'Isabella is unpredictable, Maria. Who knows? Now I must attend to my correspondence.'

For several days there was no untoward outcome of this little scene, but both Christopher and Maria were watchful, and as far as was possible, extremely solicitous of Isabella's welfare.

What neither of them anticipated was a communication addressed personally to Madame Isabella Weber, via the Diplomatic Courier and the Ruritanian *Chargé d'Affaires* in London, inviting the said Madame Isabella Weber to assist at the Coronation of His Highness the Grand Duke Karl Ernst Maria Michael Frederik, on the First Day of January in the year One Thousand Eight Hundred and Ninety-eight, in the Cathedral of St Michael. A seat of Honour would be set for her.

This missive, headed by the Arms of the Royal House of Ruritania in purple and gold, was subscribed by the Lord Chamberlain in writing which resembled the grotesque meanderings of a black beetle.

Isabella read this unexpected communication with the greatest interest. A sudden exclamation as she reached the penultimate line brought Christopher to her side. 'What is it, Isabella?'

Speechless, she held the paper towards him. A little way below the signature, as though it were reluctant to acknowledge the relationship, there appeared, in the same senescent script, a kind of *post scriptum.*

'*I beg to remain Your Highness's faithful servant.*' Christopher read it aloud in bewilderment.

When at last they had both recovered from the shock of this strange message, Isabella said:

'*Your Highness . . . faithful servant . . . .* What can he mean, Christopher?'

She turned the paper right and left, before and behind, seeking for some answer to the riddle. There was none.

'What can he mean, Christopher?'

'I do not know. Perhaps it is a mistake. Perhaps you have received a message intended for someone else. It is very intriguing, nevertheless. It can't be intended for you, Isabella. The Lord Chamberlain must be past it.'

The mystery was no nearer to a solution when, some days later, a further missive arrived by the same channel, inviting Madame to sing the leading role in an opera to be mounted in the presence of His Highness and guests in the evening of his Coronation Ceremony. The reaction to this was somewhat different.

The invitation, subscribed by the four semi-aristocratic members of the Opera Committee, began:

'*Madame Isabella Weber.*

In homage to your divine voice . . . .'

Isabella exploded in fury.

'My divine voice . . . . What hypocrites they are. If they want me to sing, why do they not say so immediately? They have not heard my voice for twenty years. They have no idea whether my voice is divine or not. And this — '*My beautiful presence* . . . .' Faugh. It sickens me. And what do they suggest? *La Sonnambula; Lucia di Lammermoor* . . . . What do they think I am, back there in Strelsau? Am I to sing for my supper all over again? Me, their ''backyard hen''?'

'You are too anxious to think the worst, Isabella, and that was a very long time ago. You are, first, recognised internationally as supreme. In the second place, you have sung, not too long ago, at Munich, Dresden, and Vienna. The people of Strelsau are not fools. It is highly probable that many of them have both seen and heard you many times in those places.'

Isabella glared at him.

'After twenty years,' she stormed. 'Not a word of any description during twenty years, and now this. Now they beg me to sing for him . . . them . . . . It is not to be borne. The answer must certainly be no. You will inform them immediately, Christopher.'

'I believe that you are very wise,' Christopher considered. 'You do not have to sing in Strelsau. It cannot but do harm to revive memories which should remain memories. I do not think you should sing again in Strelsau, whatever the occasion.'

She looked at him with speculation in her eyes. Christopher became uneasy.

'Besides,' he went somewhat lamely, 'there is Lisa.'

'Ah, yes. Lisa,' said Isabella in a soft voice.

'What if she should hear of the circumstances?' He was fighting in desperation now, aware that he was losing ground.

'I can only advise you against it, Isabella. It would be folly in the extreme. Do not go, I beg of you.'

Isabella drew herself up. Her eyes flashed.

'Lisa, yes. What was the message? *Your Highness's faithful servant* . . . I will sing for them in Strelsau. But not *La Sonnambula*. Indeed not. They shall hear what I intend them to hear.'

'Isabella. You cannot do this,' was Christopher's despairing cry.

*

It is many years since she has drunk the waters of Strelsau as recommended by 'Corno di Bassetto'; many long years since she has seen snow falling so white and clean as it does in the Stanislausplatz; many years indeed since she has listened to the ripple of the water as it flows through the sombre woods, or watched with delight the sparrows quarrelling over crumbs around the base of the statue of the Grand Duke on his horse (surrounded now by seats for the benefit of those who wish to sit for a while, be it summer or winter). These things remain as they always were; but so much has gone. Dear Herr Weingartner's house, that he was so proud of, turned into a chocolate shop, and Josef himself dead these many years, his '*Weingartnerteorie*' a dusty pile of papers on some forgotten shelf. Matthild' his beloved sister and spinster by nature so compelled by circumstances to have married the Doctor, and both growing old; Marchesi gone, her place taken by her daughter Blanche, an indifferent singer, but with some success no doubt; the old king, last of the Ruritanian tyrants, dead these three months, and the new king crowned this very day with his unlovely wife, Amelia Viktoria Konstanzia Camilla, beside him. Before all, Papa Smith, her dear Papa, whose only fault was that he loved and admired his adopted daughter too well, and lacked the courage and the wit to control her pride, her arrogance, and her selfishness.

All gone, and yet not all.

She covers her face in an endeavour to shut out the crowding ghosts of the past wavering and gesticulating in the dark mirror.

'Why do you return?' they leer, advancing and retreating alternately. 'Why have you made this monstrous journey? There is nothing here for you, Isabella Weber, *dear* Madame Weber. Your time is past. Why did you come . . . come . . . come . . .?'

The ghostly voices echo in the glass prison.

She stares at them, her eyes wild, her face ravaged.

'I had not wanted to come . . . .' she whispers to that sorrowful reflection. In her despair she tries to make herself believe that this is the truth.

The ghosts deride her and jeer again; they poke and mime in the further darkness.

'She had not wanted to come,' they mouth to each other, their outlines shaking with glee. 'She had not wanted to come, yet here she is. Mark those tears, and that sad sighing.'

On the table before her the little travelling clock lends its self-important voice to the next extract from Master Mozart's 'Air' in its repertoire; out in the corridor a bell announces that there is no longer time for dallying in dressing room or band room. There is the sound of urgent feet and muted weary voices.

'*How beautiful you are*,' he had said, standing just inside the door, that same door through which she must go in a very short time to sing the last scene of all in this strange story.

'*Together we will celebrate a double anniversary* . . . .'

A very conceited young man, she had thought of him.

'*I will call upon you again* . . . *tomorrow* . . . .'

Those words. Three days later, had been his last words to her, the last time she had heard his voice, twenty years ago.

The ghosts whinney and sniffle in the mirrored darkness. They advise; they threaten; they giggle once more.

'Why do you disturb yourself, Isabella Weber? It is too late . . . . Too late . . . .'

She lifts her head and gazes steadfastly into the mirror. Why indeed should she disturb herself? There can be nothing to gain by it. Away with memories. Let the dead Past bury its dead. She outstares the ghosts. But they will not be restrained. They posture through the glass at the face shadowed before them. They mimic the scenes in which they have been severally involved, and their mimickry spreads far and wide, from city to city, portraying Isabella Weber in her moments of triumph; Isabella in Vienna, in Budapest, in Dresden, in Munich, in Milan; Isabella in London, even. Isabella, gathering love and hatred, applause and criticism, awards and honours.

'Such a restless life,' they babble, 'and all to what end?'

Such a restless life . . . Vienna. Madame Weber will appear . . . Vienna in the winter snow; Madame Weber in *Fidelio* . . .

Paris in the cold miserable winter; Madame Weber in *Faust* . . .

Moscow in snow and a suspicion of Northern Lights, and Madame Weber in *Romeo and Juliet* . . . .

And now Strelsau, and Hans the Doorkeeper in his little office. His Majesty sitting with his porcine wife in the Royal Box, and Isabella Weber, the well-known opera singer to entertain them. Madame Weber, 'Our Isabella', in *Alma Perdita*.

Isabella gives a great sigh.

'And out of all this, what?' she demands passionately of the ghosts.

They wave nebulous purses of gold, and a handful of ribbons and medals before her listless eyes, and recite in their cackling voices the meaningless jargon that goes with them.

Kammersängerin to the Imperial Court of Austria-Hungary . . .

Kammersängerin to the Court of Prussia . . .

Gold Medallist of the *Academia Reale Musicale*, Napoli . . .

Gold Medallist of the *Academy of Music, St. Petersburg* . . .

A cameo brooch from the personal collection of Her Majesty Queen Victoria . . .

From Strelsau . . . nothing.

'Isabella. Isabella Weber,' they continue to cry, wishing only to keep before their eyes that stately figure and those cold green eyes, and that dispassionate, arrogant expression which causes them to shiver in ecstasy.

'Isabella Weber. *Our Isabella.*'

Weber . . . Weber . . .

IsabellaWeber . . . Isabellabellabella Weberebereber . . . .

The applause echoes and re-echoes on the high, painted ceiling and about the dusty pillars and cornices.

*

With little intention on her part Isabella's fame had spread like a conflagration. At breakfast tables in London, at dinner tables in the cities of Europe, and at supper tables in the cities of America she was almost the sole topic of conversation. Dogs and hats were named for her — although contrary to the dictates of fashion she did not keep a dog, and seldom wore a hat. Windows in the Rue Royale, Bond Street, and the Kärntnerstrasse displayed simultaneously 'Isabella Evening Gowns', 'Isabella Tea Gowns', 'Isabella Shoes', and 'Isabella Stockings', although for all anyone knew to the contrary she herself had never appeared in any of these fashionable confections.

At superior *modistes* a certain intimate garment of stout whalebone and calico, called 'The Diva', was to be purchased by all ladies between the ages of eighteen and fifty who sought to mould themselves into a superficial likeness of the exquisite Isabella. Inserting themselves into this garment with the greatest difficulty, the unmarried would urge their shovel-handed Abigails: 'Pull, Dolores,' (or Mary, or Gretchen, or Simone, as the case might be). 'Pull, Henri,' the married ones would exclaim, hanging on to the bedpost in agony, while Henri, placing his foot in the appropriate place, would haul on the unforgiving laces with all his strength, seeming to derive some peculiar enjoyment from the exercise. It was all in vain. No corset ever built could confer even by the reduction of several inches the faintest resemblance to that perfection. *A man's dream; a woman's despair*, one waggish critic wrote of it, despairing himself of ever being able to describe Isabella's physical presence.

Thus imprisoned, the ladies continued to swoon delicately at *soirées*, while Isabella moved freely, her unique self.

Uniquely she was a rock against which the waves of adulation and admiration washed and broke harmlessly. Season after season she sang, without apparent effort, without displays of temperament, coldbloodedly. She sang as she was expected to sing, and as she was paid (at her own valuation, it must be said) to sing. When the last note was sung, her work was finished, and she left the stage. Her fellow artistes were accustomed to this eccentricity, but audiences who relished the exhibitionism of other singing ladies, who could scarcely be accused of 'singing' but made up for it in other directions, could not make head or tail of her attitude.

Her refusal to indulge the demands of her audiences in the matter of *encores*; her disdainful behaviour in the matter of curtain calls, the display of her collection box for orphaned children in the foyer of every theatre in which she sang, caused her admirers to consider deeply that she was a phenomenon beyond their understanding.

'She sings!' they exclaimed, as if they had really expected her to hang from the chandelier or leap about the tiers of boxes as mountebanks sometimes did at Astley's Circus. Some were not even sure that she was a woman. Perhaps she was just a clever machine, the like of which was quite the rage at Theatres of Magic. Perhaps she was another *Trilby*; credence was lent to this idea by the fact that she was on all occasions accompanied by a man whose left hand was encased in a black velvet glove. Yet everywhere they clapped until their pale hands became red with the contact of flesh upon flesh, and they vociferated until their own vocal cords twanged in their throats.

'*Bis. Bis*,' they cried, as the unmoving, unsmiling figure on the stage looked down into the darkness which hid them from her. When their inordinate exuberance was finished, she would continue with her particular role, but not a moment before. They were shocked, of course. No singer had ever behaved in this manner. But they were not discouraged. They considered that they could change all that; they considered that every singer, *Prima* or not, had a point of capitulation. So they continued to clap and shout, to demand encores and extended curtain calls. When the acting and the singing were done, the gentlemen of the opera-going species who were fond of themselves and lustful of her appearance, her talent, and her rumoured wealth besieged her closed retiring room. When this tactic failed they bombarded her with flowers, gifts, and invitations to dine. The flowers and gifts were invariably returned by the messenger who brought them; the invitations to dine were ignored. Yet such is the steadfastness of worshippers that season upon season, in city after city, they crowded in increasing numbers to feast their eyes on that beauty, to listen to that perfect voice.

'I shall sing like Jenny Lind.' Isabella frequently recollects her childish boast. 'The people will clap, and give me money.'

Indeed they clapped, *and* gave her money, for there was little possibility of comparison between the two nightingales.

But to what end?

The ghosts continue to poke and pry, seeking to breach some weakness in the wall she had built about herself.

'. . . *Why do you disturb yourself, Isabella Weber? What can you do? It is too late . . . too late.*'

'What shall I do?' she cries aloud to the glassy reflection, pleading for some tardy support to the resolution which has brought her from London to Strelsau, and is now like to fail her.

'What shall I do, Maria?'

'It is too late, Isabella. Far too late.'

Maria echoes unwittingly the mockery of the ghosts, but she speaks gently. 'Both Herr Christopher and I advised against returning to Strelsau. But since you are here, I myself recommend that you do your singing and depart with all speed.'

Isabella stares at her with tragic eyes.

'Without explanation? Without knowing, Maria?'

Maria shrugs her shoulders.

'What can it change, child?'

'I loved him, Maria. I loved him.'

'Much good did it do you,' Maria comments dryly.

'I love him still, God help me.'

'Much good *will* it do you. It is useless, Isabella Weber.' Maria speaks with some loss of patience. 'You are a grown woman, an internationally-known opera singer, and you are behaving like a spoilt child. If it was merely a man you wanted, you have had opportunities by the hundred.'

'How true that was,' Isabella reflects bitterly. Christopher, Andelli, Hans Richter . . . and those anonymous young men who turned pale with suppressed emotion or frustrated desire whenever she sang to them, and those who wished to make a closer, more personal acquaintance and besieged her retiring room, the door of which remained firmly closed as Maria learned a little more about life as Isabella's companion.

. . . Except, she now sometimes reflects sadly, on that one occasion, so long ago, when she too had been somewhat naive. It is possible that she has been too precipitous this evening, reflecting what she imagines are Isabella's feelings in the matter, in denying His Majesty access to Madame Weber? Who can tell what moves the human mind? Throughout the last intermission both Maria and Isabella have listened with ill-disguised alertness for the sound of purposeful footsteps and the peremptory knock on the door.

The ghosts mew and mouth in their dark emptiness.

He has not come . . . . He will not come . . . come . . . come . . . . It is too late. It is beyond reason . . . .

Isabella turns suddenly to face Maria.

'Maria.' It is a command.

'Madame?'

'Maria,' Isabella's voice is urgent.

'A pen and paper, quickly.'

'But, Madame . . . .'

'A pen, I say, and paper.'

In the mirror, behind her pale reflection, the ghosts grimace and cackle. *It is quite useless. She will not see him. We are quite sure of it.*

. . . Their outlines shimmer and waver as she puts pen to paper with a decisive gesture. They flicker into obscurity.

> *Madame Isabella Weber presents her compliments. She will receive His Majesty, alone, in her retiring room at the conclusion of the presentation.*

There is the sound of distant laughter and faint repetitive voices: *Useless . . . . It is quite useless . . . less . . . less.*

Maria's features are grim.

'It is quite useless, Isabella. You disturb yourself for no reason. What good can come of it now?'

'He has to know, Maria.'

'But there are other means of acquainting him with the situation, if acquaint him you must. This way is quite useless.'

'You think he will not now come?'

'Of a certainty he will come, Isabella, but to what purpose?'

'That remains to be seen, Maria. Meanwhile, this message must be given into his hand now, before the curtain rises.' Her voice takes on a sharper note. 'Attend to it. There is little enough time.'

Maria curtseyed.
'Yes, Madame.'

*

In the pit the timpanist makes tentative adjustments to his instruments. This always has to be done *in situ*, since he cannot carry them away with him at each interval. As he goes through these motions, he yawns, he scratches himself beneath the arms, he eases his too-tight cravat. He is bored, to put it bluntly. Scarcely one of the orchestral aristocracy, he is denied the ecstasy of listening to the noises they produce from the scraping of the gut or the blowing of wind through wood or metal tubes. There is no opportunity for interpretation in the beating of his drums. He follows the printed directions on the paper before him, and attends to the conductor's signals when that is necessary, and that is all. Whatever is wanted in the way of style, feeling or emotion has been indicated by the composer. Lacking somewhat in imagination, he presumes that Madame Weber follows the same procedure. It has never occurred to him that singing as Madame Weber sings is something more than the ability to produce the right sound at the right time, perhaps for the reason that he has heard so many in that same theatre who could do neither. What does disturb him, given his vague thoughts concerning social unfairness, is the difference in status. He has passed his productive life as a lowly member of a provincial orchestra, while she, born by the grace of God, in the same provincial city, it is said, seems to hold pride of place as the most important citizen. Compared with her much advertised presence, the coronation of The Grand Duke Karl Ernst Maria Michael Frederik has already all the effect of an interesting sideshow.

Everywhere it is *Madame Weber, the Prima Donna . . . Assoluta. Madame Weber to sing Cordelia . . . Madame Weber to sing* 'Alma Perdita.' It is as though the city were preparing for a Roman Holiday.

There were of course the dissenters, those who did not believe in according with the *vox populi*. In coffee shop and in salon other rumours rapidly gained substance, fuelled by gossip of such volume that the city was shocked and its allegiance sharply divided.

Madame Weber to be a personal guest of the Duke, gossip said.

Opera singer at the Castle . . .

and:

Madame Isabella Weber at the Coronation . . . .

The townsfolk were appalled and outraged. 'An opera singer invited to the Coronation . . . . What were the Duke and the Lord Chancellor thinking of? In the Cathedral. A place of Honour . . . . Monstrous, an outrage. Obviously now is the time to show how democratic the country is. Now is the time to invite also the flower girls and the fiaker-millis . . . .'

The timpanist's wife, like other housewives of Strelsau, was a born gossip. Sometimes the timpanist fancied that what she lacked of fact, she supplied in the form of fiction. He could not otherwise account in any way for her continual supply of malicious tidbits. As a general guide to sanity he made a practice of not believing everything she told him.

'But it is true,' she insisted. 'I had it from Lisa, who had it from Berthe. And everyone knows that Berthe's second cousin is the third groom in the Royal Stables.'

'And I suppose His Highness consulted his third groom before making arrangements for his Coronation?'

The timpanist, rummaging in the armoire for a clean shirt, was quite sarcastic.

'Further,' continued Madame Timpanist, quite unabashed, 'it is said by some that she will not be present at the ceremony . . . .'

This was a surprising statement. The timpanist was quite interested.

'It is said that she has refused,' his wife affirmed. 'There, now. What a disgrace. What an insult to His Highness.'

She mimicked Madame Weber refusing: *Madame Weber, the Opera singer, begs to refuse His Highness The Grand Duke's invitation to assist at his Coronation.*

Somewhat perturbed at this possible turn of events, even though it was none of his business, the timpanist put on his clean shirt and went off to the rehearsal of *Alma Perdita*.

If it was none of the timpanist's business, it appeared to be the business of everyone else in the city. Never had there been such an opportunity for unbridled gossip. The scarcity of actual knowledge led to an increased volume of rumour. From the lowest to the highest ranks of Society the insult — for it could be considered as nothing less — to His Highness the Grand Duke, to Her Highness the Grand Duchess and, not least, to the Lord Chamberlain (now a spindly-legged old man whose white hair and beard were constantly agitated by the troubles of his world) was almost the sole topic of conversation. An invitation from the Lord Chamberlain was in effect a command from His

Highness himself. To ignore it, would bring down upon the head of the recusant the severest disapprobation, if not worse. How then could this opera singer — for although Madame Weber might be *la Assoluta, la Superba*, and worshipped throughout the world, she was still only an opera singer, strutting and fretting her hour upon the stage, and if rumour were further to be credited, a gutter brat fortunate enough to have been endowed with a good voice — have the arrogance to refuse such a command?

The curious citizens of Strelsau waited on events.

They had not long to wait. In the early morning of the first day of the New Year, as flags were broken at mastheads, and bells began to clash and clang in wild abandon to announce the forty-first birthday of His Highness The Grand Duke, to be crowned King in a short time, and workmen were setting up table for the distribution of free food to the citizens, and putting up hogsheads of wine to be broached during the coming festivities, a carriage drew up before the *portière* of Madame Weber's modest lodging. The solitary driver descended from his seat, tethered his horses, and with some condescension beat upon the door with his whip handle. There was time enough before he was admitted for an amorphous drift of grey faces to note that although the plain black carriage bore no Arms, the condescending young man wore the Grand Duke's livery. He was without doubt from the Castle. The grey faces moved and turned, silent and ghostly in the raw morning air. At the sound of footsteps from within, and the screaming of the great door as it was opened, the faces came together in a mass; from their open mouths came a whisper, a sibilant susurration of sound in which the name 'Isabella' was clearly audible. Looking neither to right nor left, she took her place in the carriage with dignity, and arranged her furs about her with some deliberation. The young man further placed a great bear skin about her for her further comfort. It was noticed by the curious eyes in their ghostly sockets that in spite of her dignity and her composure, the beauty of Madame Weber's features was a little marred by some inner tension or conflict of the spirit. The young man, sulky and grim-faced, untethered his animals, mounted his seat and gathered up the reins. He cracked his whip, and with some preliminary prancing and snorting on the part of the horses, as though it were quite the thing to do, the equipage moved off.

But not in the direction of the Cathedral.

How well she remembered it.

The door was already open, as though she were an expected guest, as the carriage stopped with a skid of wheels and a jangle of harness before it. On the step stood a sturdy young man of the peasant type, his broad face brown from constant exposure to the weather, his eyes wide at the scene which presented itself. He noted the imperious beauty of the face, the red bronze of the hair under the travelling hat worn for once as a concession to the unreliability of the weather, the richness of the furs which enwrapped her. He noted also the plainness of the carriage contrasted with the splendour of the animals that drew it, their accoutrements, and the recognisable livery of the driver.

He bowed low as Isabella stepped down.

'My lady . . . .' he greeted her.

'I would speak with the Herr *Wirt*,' Isabella informed him.

He bowed again.

'I have that honour, my lady.'

Isabella considered him for a moment.

'Time passes,' she mused. 'I remember an older man, though somewhat of your look . . . .'

'Indeed, my lady? My father, perhaps, an old retainer of the Grand Duke's Household, established in this Inn by the Duke himself that his — my father, you understand — old age might be spent the more comfortably.'

A smile curved Isabella's lips.

'And for other . . . purposes too, no doubt,' she murmured.

'My lady . . .?'

'No matter, Herr *Wirt*. He was paid well enough to be discreet. I have no doubt he was a good and loyal old man.'

'Alas, dead these many years. Unaccountably I still receive his provision, although the Duke himself I have not seen . . . .' he paused.

'For twenty years, to the day,' said Isabella impulsively.

He was clearly surprised.

'How should you know that, my lady?'

His eyes searched her face intently. Then it was as though a cloud had passed away from before the sun. He smacked the fist of one hand into the palm of the other.

'By God's Death,' he exclaimed. 'Now I remember. I am a dolt. But it has been a long time . . . .'

His next words were scarcely such as she expected to hear.

'Forgive me, Your Highness,' he begged, and repeated: 'It has been a long time.'

A sudden shiver of apprehension ran through Isabella. Again that phrase. *Your Highness's faithful servant* . . . . Had the world gone mad?

'What was that you said,' she demanded sharply.

The inn-keeper's face reflected his bewilderment.

'Your Highness . . .' he began.

'Yes, yes,' Isabella stopped him. 'Why do you use that title to me? How and when did you learn it?'

He was puzzled by her reaction to what his memory told him was a correct mode of address.

'How, and when? I cannot say, precisely. My father told me, binding me to secrecy . . . I have grown with the knowledge of it, but I have had little occasion to think of it. As to why I use it . . . .'

As he paused to think on the problem, Isabella indicated with a sharp glance the lackey who, while seemingly busy with his horses, was earnestly pretending not to listen to these exchanges.

'Consider your words well, Herr *Wirt*,' Isabella advised the landlord, 'and your silence may indeed be golden. Meanwhile you may continue to address me as 'Madame'. Such an honorific is usual, is it not?'

'Very well . . . Madame, I understand.'

But it was clear that he did not.

'We may well speak of the other matters on another occasion,' she said.

He inclined his head.

'Now,' said Isabella, 'since I have so little time — and because this is a foolish and sentimental journey — I require the use of your best chamber for the space of an hour. I am somewhat wearied, and would rest. Attend to the young man out there, and his horses. Give him food and drink (at His Highness The Grand Duke's expense, if that is possible). You will find the young man sullen, with a tendency to be unco-operative. Ignore it. Above all, do not engage him in conversation.'

'Very well, your . . . Madame. It shall be attended to immediately.'

Isabella smiled upon him, adding greatly to his confusion.

'Your understanding does you credit, Herr *Wirt*,' she told him.

Neither more nor less than an hour later Isabella mounted the carriage once again and her comfort was attended to, with noticeably a little more respect, by the sullen young man.

'To Strelsau,' Isabella gave her orders, 'with such speed as is compatible with safety for your animals, yourself, and me. There you may take me to the house of the Surgeon Officer of the Opera and thereafter to the Opera itself — the Artistes' Entrance. Go now.'

In his corner of the wings Christopher checks the time. The piece is running late, and lateness is not one of Isabella's weaknesses. Christopher supposes that because of the occasion some variation from absolute routine is acceptable. He is more concerned at this moment with Isabella's health rather than her singing. It has been a long and gruelling performance, and the Doctor's opinion has not been at all encouraging. There is half an hour to go to the final curtain. After which, it may be, the whole situation may finally be resolved one way or another. To what end, he cannot surmise.

Backstage there is a general atmosphere of weariness; in the corridors, deserted now, and silent, a silence broken only by the reflected and shattered sounds of music creeping round corners from the stage. There are no joyous voices, no patter of hurrying feet; there is no colourful swirl of dresses where, a relatively short time ago, there was life and laughter, even love.

On stage, however, the picture is somewhat different. Here there is violent and colourful movement as the chorus of peasant maids and youths express their horror at the tale unfolding before them. Cordelia, her hands outstretched, endeavours to intervene in the quarrel of which she is the cause, a quarrel between hated aggressor and valiant protector. The music rises and falls with the cut and thrust of the duelling rivals; it is pointed by the clash of blades as the pair perform their manic dance in the confined space. Their heavy breathing gives an air of verisimilitude to their histrionic efforts; it is audible throughout the auditorium. The music and the pace grow faster, until the innkeeper's son, the young Rudolpho in love with Cordelia, skilful enough with the sword, but the inferior of the sophisticated Duke, slips and falls. He is pierced by the Duke's point, and lies dying. A great cry goes up from the Chorus.

Putting up his blade the Duke pleads with Cordelia. He asks that she should forgive him, that she should pity him for the madness of his love for her. He will do anything, even marry her, do anything she wishes to gain her love.

She rejects him. In a soaring condemnation she catalogues his arrogance, his egotism, his self-importance, his cynicism, his unprincipled shameless behaviour towards those unable to defend themselves. That beautiful voice, with all the force of outraged maidenhood behind it, pierces him more woundingly than the sharpest sword could do. He draws back from the minatory pointing finger, the distraught vengeful features. His face grows grim and dark. He tells her that if she continues to reject him he will fall upon his own sword. He tells her, in soaring tenor voice, that he cannot live without her.

She scorns him; she laughs in his face, a scorning unbelievable laugh.

He approaches her with dark determination. In the background the assembled villagers gyrate, rotate, re-assemble, and withdraw to the confines of the stage, expressing in pianissimo song their bewilderment at the situation. Now the music enters upon a variation in a minor key of Cordelia's aria from Act Two — *Non mi tentar*'. The villagers vocalize in unison upon this theme while Cordelia, still singing her rejection of the wicked Duke, bends in simulated sorrow over the body of Rudolpho.

This is my love, dead he lies at your feet, dead by your vile blade.

The villainous Duke pulls her roughly to her feet.

Not so, he replies in anger. *I am your love. You shall be mine.* And he embraces her crudely, roughly.

She struggles to repel him.

What brutality. What villainy, the Chorus intones on a rising note. The great auditorium is silent, the audience within it temporarily stunned by the vehemence manifest in the exchanges between the two principals.

'How realistic it is,' is the thought uppermost in almost everyone's mind.

The musicians, weary but elated, play as though their lives depend on it, urged on by Andelli's flying baton.

Above the *fortissimo* the voice of Cordelia rises in desperation, clear, penetrating . . . .

Why must I suffer in this way? Would that I were dead . . . . Oh Thou, God, who watchest over all Thy children, let me die . . . .

There is a sudden sob in the music; a dying fall, a heartbreaking diminuendo and pianissimo.

Let me die. Fatemi morire, Dio Salvatore . . . .

With a last despairing struggle she breaks free from the enclosing brutal arms and stands apart, her breast heaving. Andelli nurses his musicians gently, quietly, to the point at which Cordelia begins her last aria. The sound from that twisted beautiful mouth is scarcely audible, but gains in strength as the aria goes on.

Madre Santa, Virgine pietosa . . . . It is the cry of a lost soul to the Great Mother. To hear it even today brings tears to the eyes.

The Ladies-in-Waiting fumble for the inadequate pieces of lawn; the gentlemen of the party have sudden difficulty with their eyesight, attention to which involves the use of larger squares of silk. Her Majesty Amelia Viktoria Konstanzia Camilla, having finished her tally of bonbons, is rather at a loss for something to be interested in. It is sad that her acquaintance with the Italian language is so slight. She wriggles in her chair, and fans her vapid face. She looks slyly round at those members of the audience it is possible to distinguish in the reflected light from the stage. From time to time she considers more seriously the handsome actor who lately had his arms about Cordelia, and sighs deeply. But she makes no effort to understand the operatic situation. To her it is merely a tiresome entertainment to round off a really tiresome day. After all, her husband has been King in all but name for a number of years.

Her newly-crowned husband — King of Ruritania — sits unmoving by her side, as he had sat, albeit alone, watching the original performance some twenty years ago. Lusty he was then, in the strength of his youth, and greatly taken with the new singer everyone was already talking about.

How beautiful you are, and how wonderfully you sing.

The ghosts gibbered and mimicked about his memory. He thrust them angrily aside. Singing was now quite the last thing he wished to think of. But they would not be put away: *How beautiful you are . . . . And how beautiful she still is.*

Your birthday, and mine. We must celebrate more intimately, Fräulein Weber.

Again he brushes aside the ghosts of Time Past. As he contemplates that tragic figure acting its little false sentiments in the limelight, his eyes are somewhat softer. *She* has returned, even though the price of that return has appalled him, as it has the Directors of the Opera. A king's ransom, no less . . . .

He feels that the problem, the solution to which has haunted him for twenty years, is about to be solved, explained. But to what end?

To what end . . . end . . . end . . . . The resurgent ghosts mimicked him, and laughed as they mimicked. *It is too late . . . Isabellabellabella . . . .*

To what end indeed?

For twenty years, under threat from his tyrant father, of summary and condign punishment should he attempt to see her again, or engage in correspondence with her, what knowledge he had had of her had been gleaned from reports of her progress in newspapers and gossip between the musicians with whom he had once delighted to associate. When, under the same dour threat, he had unwillingly married the distasteful Princess of Zell, and had equally unwillingly assumed the problems and duties of future kingship, it was far too late. The urgency died, as is the way with young men in love. But the memory remained.

And now, she has returned. He feels his heart swelling again. He steals a sidelong glance at his Queen Consort as she sits lumpily staring with unperceiving eyes at the invisible fourth wall, beyond which Cordelia is at this moment fighting to free herself from the evil embrace of ravening lust. Her Majesty is not one whit moved by the terrible scene, nor by the heartrending music. Her only reaction is to direct a similar sidelong glance at her Lord and Master, a sardonic smile hovering upon her lips. She has naturally observed the envelope borne on a silver salver and given into his hands by a theatre lackey at the end of the intermission. Its perfume is still pervasive. She is agog with a curiosity she knows will never be satisfied.

It is as well, or so reflects His Majesty, that the little secret of his Inn beyond the hills was known to so few trusted . . . *friends*? Yet someone who was not quite to be trusted must also have known, someone who must have been very cognisant of his movements. Yet never had any word been spoken, no questions asked, no recrimination voiced. How inexplicable it was then that so soon after that impulsive gesture he had been confined to the castle by his father's orders, and Isabella had disappeared from Strelsau. There was no solution to the problem. All that was left was memory and longing for what might have been.

Our Anniversary . . . . We must celebrate more intimately . . . .

And so they had. The ghosts lay out before him the flickering pictures; they note the warmth that creeps into his frozen heart as he drives his sleigh in recall across the snow, across the bright morning valley to the tree-scattered uplands, through the further dark forest to the inn on the ledge of the mountain. They remind him how his heart had sung with joy as he had driven, keeping time with the beat of the horse's hooves, and the hiss of the runners on the snow. The ghosts snicker over his boyish boasting as he told her of his prowess in the hunting field, and they giggle as he declared over and over how beautiful she is and how wonderfully she sings.

He recalls pressing her with the exquisite food and wine that his inn boasts; he recalls his enthusiastic description of her beauty and her talent to mine host, his trusted servant; he recalls, with a momentary pang of jealousy, the obvious worship of the innkeeper's son for her . . . . He and she were, for such a short space of time, two happy children, completely enslaved by each other. Just when youthful arrogance, charming insolence, and boastfulness gave place to humility and worship he was not aware.

Suddenly, thinking on these things, his heart ached, and he felt afraid.

His heart aches again, he still fears what he does not know. He is, in spite of the regal trappings of his daily life, the press of self-important people about him at all times, a lonely man. There is no one who can help him. *She* has returned, but they are still separated by the length of the auditorium in space and twenty years in time.

He sits in his royal box, lonely, friendless, a King.

Down there on the stage, in the full limelight, Cordelia struggles frantically to free herself from her brutal pursuer. He approaches her once more, but she flies from him across the stage, through the ranks of the Chorus, to the great gate of the Nunnery. She beats with her fists upon the wicket, while the villagers intone, *sotto voce*, the melody of her last aria. There is no answer to her beating, and the wicked Duke draws nearer. Looking wildly around, she perceives the dangling bell-rope. She seizes it and pulls . . . .

The percussionist beats one stroke upon his metal tube. It is his hour of triumph. Upon the stroke, the wicket opens, and the Mother Superior steps forth. As she does so, framed in the same green and gold light that bathes Cordelia, it is to be seen that Isabella puts her hand to her heart, staggers a little, and falls to the ground, where she lies unmoving.

The music decreases in intensity, falling to its closing chords, and dies away.

The curtains fall together with an audible swish. There is a pause, and the audience breaks into a furore of applause that surpasses any other such demonstration of the evening. They clap without ceasing. They call her name until the whole building echoes and re-echoes . . .*Isabella . . . Isabellabella . . . Isabella Weberebereber . . . .*

So loud is the applause that even the pigeons roosting on the cold cornices outside are disturbed.

Isabella . . . they call.

There is no reply. The curtains remain undisturbed apart from a slight quivering as perhaps someone passes behind them.

The audience continues to demonstrate its appreciation; the applause increases in volume; the sound of clapping ebbs and flows, the sound of waves on a rocky shore. Increasingly they call her name. Such a performance, the like of which has never been seen in Strelsau, deserves every mark of appreciation they can give. She is, after all, 'Our Isabella'. When, however, the ovation begins to lose its force, as indeed it must, Andelli raises a weary baton, and the orchestra strikes up a less than enthusiastic rendition of the National Hymn of Ruritania. Those of the audience who are still in their seats rise in some confusion. All eyes turn towards the Royal Box. His Majesty is not to be seen;

only Her Majesty Amelia Viktoria Konstanzia Camilla stands there, a foolish smile on her plump, vapid face, waving her hand languidly at the glittering ranks below. The Ladies and Gentlemen-in-Waiting continue to wait patiently until she is satisfied that she has milked the fountain of admiration dry. They retire into the dimness behind her, and yawn discreetly into their grubby white gloves. Her Majesty gives one final wave, and turns away. As she turns, there is a smile of self-satisfaction on her face. It is the first real sign of life she has shown all the evening.

The mathematical rows of the auditorium break into amorphous moving knots of light and colour; it is as if an apprentice choreographer were trying out his ideas for a new scenic effect. They turn, they twist, they move; they form into long lines . . . . But there is no music to the dance, only the faint muffled roar that is the sound of many people gathered together. It is a sad noise.

The auditorium becomes gradually void, a vast echoing emptiness in which the only discernible movement is that of the great candelabrum swinging, almost imperceptibly, in the gentle breeze caused by the opening of many doors. As it swings it illuminates fitfully the debris that inevitably remains wherever people have gathered together, not least at the opera: a broken fan discarded by its fair owner; a silver button with a morsel of thread still attached, and not to be missed until much later; posies fallen from negligent laps, trampled upon, crushed beyond recognition; programmes fallen and forgotten under the seats — valueless valuable specimens of the caring printer's craft, embalming in purple and gold an occasion which from this time on can only be a memory. The power of such music as has been heard this evening cannot be enshrined in purple and gold. The outward trappings are unimportant. The importance lies in its effect on the heart, and only the heart can understand that which comes from the heart.

So they pass, bewigged, bejewelled, beribboned into the *foyer*, suddenly conscious of the biting cold, and the darkness of the night. They make their way down the carpeted steps, to stand beneath the canopy until their carriages should be called, the Ambassadors, the Ministers-Plenipotentiary, the Consuls, the Princes of greater and lesser degree from comic lands beyond the boundaries of this realm of Ruritania, suddenly of importance for the space of a morning and an evening. Proudly and self-consciously they had mounted those same steps a handful of hours ago; now they were somewhat saddened by what, after all, was an all-too-familiar human tale. As they waited, several of the gentlemen put a coin in Isabella's Box for Children.

'Madame Weber has excelled this evening. Did you ever hear such an exquisite voice?' they say as they bid each other the sweetest of dreams.

'A pity that on this evening of all evenings we were not granted at least one last appearance . . . . Still, she makes her own rules, we suppose. She has always made her own rules, I understand . . . .'

'Arrogance, my dear. I understand that she allows herself little contact with us poor mortals.'

'Arrogance indeed. One would almost think she were Royalty itself.'

Shivering, they attend their cold carriages.

The swinging pendulum of the great candelabra swings ever more slowly. The multitudinous faint shadows it creates poke and peer a little less impertinently. The ghosts grimace; they laugh inaudibly; they run back into their corners and disappear.

Their voices can be distinguished only faintly in the cold empty House: *It is too late . . . too late . . . . She cannot see him . . . .*

The lights are extinguished, and a dark stillness descends on the dusty auditorium. The pendulum has stopped.

From across the snow-covered square the bell in the tower of the Cathedral of St. Michael tolls a single booming stroke.

It is twelve-fifteen, and a cold night.

THE END

The programme of the Opera

THE ROYAL OPERA HOUSE
STRELSAU

*

A

REPRESENTATION

ON

THE OCCASION

OF THE CORONATION

OF

KARL ERNST MARIA MICHAEL FREDERIK

HEREDITARY GRAND DUKE OF RURITANIA

*

On the First Day of January 1897
at 7.45 p.m.

*

Lyrical Season 1896–1897

ALMA PERDITA

Opera

in three acts

*

The Libretto by
Signor GIACOMO GIACOMINI

*

The Music by
Signor RUFINO ANDELLI

*

The First Representation of the Opera was at Strelsau on the First Day of January 1877 at the Royal Opera House.

CHARACTERS and INTERPRETERS

*

CORDELIA, a young lady of aristocratic birth, now an orphan.	Madame ISABELLA WEBER
MARELLA, Cordelia's Maid	Mademoiselle LISA WEBER
THE DUKE, Cordelia's Betrayer	Signor GIOVANNI TORTONI
THE ABBOT	Signor DEMETRIO APOSTOLOS
THE INNKEEPER	Signor ENRICO MENTONI
RUDOLFO, his son	Signor ANGELO GIACOMOTTI
THE ABBESS	Madame BIANCA ALVI

Villagers, Huntsmen, soldiers.

*

TIME AND PLACE:
The middle of the eighteenth century. In the mountains near Perugia.

*

ACT I

Before the Church of the Holy Sepulchre

Evening. Strelsau.

ACT II

Scene i

The market place in the village of Buonasera.

The afternoon of the next day.

Strelsau — Vienna — Strelsau.

Scene ii

The Inn of the Cat-in-Boots in the same village.

The evening of the next day.

ACT III

Scene i

Outside the Nunnery of the Madonna dell'Angeli.

Two days later.

London — Strelsau

Scene ii

The same place and time.

MADAME ISABELLA WEBER'S ARIAS

*

In ACT I

Sono pellegrina

Senz' amore, senza vita

Peccavi

In ACT II

Si Signor, son' Orfana

Non mi tentar

In ACT III

Madre santa, Virgine Pietosa

THE ARGUMENT

Cordelia, a young lady of high birth, is betrayed by the Duke. When the fact becomes known, he denies all complicity, since he is already betrothed. When Cordelia insists that he should marry her, he rejects her, even going so far to accuse her of using witchcraft to make him her lover. She denies this accusation, but is forced to flee from fear of the superstitious villagers.

She becomes an outcast and a wanderer, with only her faithful maid for company. Together they beg their way like gypsies from village to village. Cordelia vows to spend her life searching for the meaning of 'Truth', to replace the love that has betrayed her.

Coming upon the Inn of the Cat-in-Boots, Cordelia is treated with compassion by the Innkeeper and his wife, and is given freely a room for herself and her companion until they are able to resolve their situation.

The son of the Innkeeper falls in love with Cordelia. She in her turn finds him attractive, but tells him she cannot love him in return. He vows to serve her until she can.

While she is at the Inn the wicked Duke arrives, thirsty from hunting. Finding Cordelia there, his passion is aroused once more. She will have nothing to do with him, although he says he will now marry her. He attempts to take her in his arms. She thrusts him aside and flees into the Inn. He follows her. The Innkeeper's son enters, sums up the situation, and accuses the Duke of being a libertine. The Duke, frustrated and insulted draws his sword, and challenges the young man to a duel, mortally wounding him.

Cordelia, witnessing the duel, is horrified. Once again she recites the catalogue of the Duke's crimes. He repeats his offer of marriage. He pleads for her pity, in vain. She flees from him once again. He pursues her in anger, and before the Nunnery of the Madonna del'Angeli they have their final encounter.

He seizes her roughly. They struggle. In the struggle Cordelia manages to draw his dagger from its sheath, and stabs him.

Running to the nunnery gate, she pulls frantically on the bell rope, and while doing so collapses at the feet of the Mother Abbess, who is at that moment coming out.